I0605635

ALONG LAKE MICHIGAN

ALSO BY MICHAEL SCHUMACHER
PUBLISHED BY THE UNIVERSITY OF MINNESOTA PRESS

Mighty Fitz: The Sinking of the Edmund Fitzgerald

November's Fury: The Deadly Great Lakes Hurricane of 1913

Too Much Sea for Their Decks: Shipwrecks of Minnesota's North Shore and Isle Royale

Torn in Two: The Sinking of the Daniel J. Morrell *and One Man's Survival on the Open Sea*

The Trial of the Edmund Fitzgerald*: Eyewitness Accounts from the U.S. Coast Guard Hearings*

ALONG LAKE MICHIGAN

SHIPWRECK STORIES OF LIFE AND LOSS

Michael Schumacher

University of Minnesota Press
Minneapolis
London

The University of Minnesota Press regrets that Michael Schumacher did not review the final editing or proofs of this volume.

Published by the University of Minnesota Press
111 Third Avenue South, Suite 290
Minneapolis, MN 55401-2520
http://www.upress.umn.edu

ISBN 978-1-5179-1677-0 (hc)
ISBN 978-1-5179-2122-4 (pb)

Library of Congress record available at https://lccn.loc.gov/2025001945

Printed in the United States of America on acid-free paper

The University of Minnesota is an equal-opportunity educator and employer.

33 32 31 30 29 28 27 26 25 10 9 8 7 6 5 4 3 2 1

For all those who lived, worked,
and died on Great Lakes waters

CONTENTS

INTRODUCTION

THE BEST WAY to gain perspective about Lake Michigan's size may be to view the lake by air. You already know that it's a large body of water, that next to Lake Superior it contains the most water of any lake in the United States, and that only Lakes Superior and Huron boast more surface area. It runs north to south, and it's deep and cold.

You can get most of this physical perspective by studying a globe or map, or by reading about the lake online. You'll learn that Lake Michigan is the only one of the Great Lakes entirely contained in the United States, and you'll learn about how it was carved out by glaciers. The more you read, the more you'll learn that it has a long, colorful history, whether you're considering the Native peoples who lived near the shoreline, the French explorers who were the first white men to see it, or the development of the land.

I've lived near Lake Michigan for all but the first years of my life. I used to beg my father to take me there. I studied the lake's geographic and social history in school. For my eighth-grade graduation trip, we took a ferry across the lake from Wisconsin to Michigan and back. As an adult, I've visited the lake every time I had the chance, and for seven years, when I lived on a small boat harbor and could stand on the deck and look out over the water, I saw how a big nor'easter could blow in and, even close to shore, whip the waves into a frenzy.

I learned to love the lake and respect its power, but it wasn't until the first time I flew over the lake that I truly appreciated its enormity. Sailing across its expanse was one thing. I was young, and the boat I was on, the *Milwaukee Clipper,* moved at a slow pace. We were in no hurry, so it didn't matter. Flying over it was different. I'd look out the window and there was blue water as far as I could see, and every so often I would see a freighter, more than the length of three football fields, chugging along a couple thousand feet below and looking like a bathtub toy that my younger siblings might play with. It took a while to see land again, and (unlike boats) planes moved at a pretty good clip. I was impressed. I might have said something to the person seated next to me. I don't recall. That was a long time ago.

I mention all this because I believe it is important in a book of this nature. Lake Michigan holds more shipwrecks than the four other Great Lakes combined, and its sheer size has a lot to do with it. Sailors could be out a long time without setting foot on land. This was true even when the weather was good and the sailing easy. When the weather turned nasty, especially late in the year, the lake was not a pleasant place to be. All of the wrecks depicted in this book occurred before GPS and, in most cases, radar; skippers relied on information that was quite often spotty, and encountering a vicious storm was dangerous. In the right circumstances, massive waves would build up as they moved down the length of the lake; on some occasions, these waves, the height of buildings on land, could roll over a boat's deck and threaten to capsize it. If you're in the middle of the lake, what do you do?

As two losses by fire in this book show, just a few miles can mean the difference between a safe return to shore and the loss of hundreds of lives. Both boats were wooden, and both burned to the waterline before help could arrive. The first shipwreck described in this book, the *Phoenix,* burned just a few miles from Sheboygan, Wisconsin—not an ideal port at the time, but one that might have mattered if the lake had been just a little smaller.

THE FIRST KNOWN GREAT LAKES SHIPWRECK, at least in commercial shipping, might have occurred on Lake Michigan. (I say *might* because the vessel has not been located.) Launched in 1679, the *Griffon,* constructed near the Niagara River, was a French-built vessel intended for trade on the Great Lakes. The first Europeans to see what is now known as Lake Michigan, French explorers traded with Native peoples, who were accustomed to plying their trade on canoes that rarely strayed from the safety of the lake's shoreline. They were awestruck by the large two-masted vessel with a unique masthead. Operated by La Salle, the *Griffon* was the first of its kind on the lakes, and the trading with the Native Americans was successful, but its life was very brief. Laden with six tons of furs destined for the East Coast, the *Griffon* set sail from near Green Bay, headed for the Straits of Mackinac on September 18; La Salle remained behind to explore the area with a team of his countrymen. No one knows exactly when or where the *Griffon* sank, but it is known that she encountered a savage storm that sank the boat and killed her six-man crew.

Commercial traffic on Lake Michigan evolved over the following centuries. The boats grew and became much stronger, reflective of the needs and demands of those paying for their services. The Midwest developed, with a large number of European immigrants setting in and building villages near the lake. The boats delivering package

Launch of the *Griffon*

goods and wheat to these communities were wind-driven. Wooden schooners dominated, though warships, commanded by the British, held the Great Lakes in check during the War of 1812.

The schooners learned quickly how temperamental the lakes could be. They depended on wind, of course, but the wind could be a mortal enemy when there was too much or too little of it. By the mid-nineteenth century, shipbuilders were working to eliminate wind from the shipping equation, and with the advent of steam-driven engines came the evolution of boat design. The early boilers were a hazard to wooden ships; fires brought down more vessels than commercial shipping companies could address. Still, with an increasing demand for such goods as lumber, metal, and wheat, steamers were a necessity. The answer was to build vessels from metal, rather than wood.

The onset of iron and steel hulls set design change into rapid motion. Schooners were the first victims, eventually being converted into consort vessels towed by the larger craft but invaluable for their ability to hold cargo. Designers changed the look of the freighters,

The *Carl D. Bradley*

turning the large midship superstructures into fore and aft deckhouses, opening up more cargo room in the center of the vessels. Freighters, large to begin with, expanded to behemoths capable of transporting previously unimaginable tonnage. For about a quarter century, spanning the end of the nineteenth century and the beginning of the twentieth century, a peculiar-looking vessel, commonly referred to as a whaleback, enjoyed popularity until the unusual design created safety issues. This book follows these developments, making it an abbreviated history of commercial shipping on the lake.

What did not change were the hazards that sank the vessels. The perpetual disagreements over safety and commercial profit remained consistent over the years. Safety was always a concern—nobody wanted a vessel to sink or men to perish—but safety measures conflicted with profit margins. As this book points out, the simple inclusion of lifeboats was an issue that was addressed and amended over the years. The early vessels barely carried enough lifeboats for the crew members, and passenger liners had nowhere near enough to accommodate everyone. The loss of the *Titanic* engendered immediate discussion about the need for lifeboats to carry everyone on a boat. But the subsequent laws requiring this actually contributed to the greatest loss of life in Great Lakes history when, shortly after the sinking of the *Titanic*, the *Eastland*, a passenger vessel filled to capacity, rolled over in the Chicago River. To comply with the new lifeboat requirements, a new level of lifeboats had been added to the *Eastland*, upsetting the boat's metacentric height and making her top-heavy.

Safety concerns, commerce, and bad sailing conditions were an awful mixture. Shipping company officials paid lip service to the principle that the captain made the decision about sailing in rough weather, and while this is true in theory, it is also known that in late fall (the time of year that was roughest on vessels, when the skies darkened and storm systems settled around the lakes) shipping firms tried mightily to squeeze in a final run or two. Depending on what had transpired over the season, those final couple of shipments could spell the difference between profit and loss. Captains were aware of this, and if they weren't, they were reminded of it. That, along with the arrogant belief that their boats weren't susceptible to the hazards that brought down other boats, led to disaster.

Heavy-weather captains, as they were called, would take their boats out in any weather or sailing conditions. They were concerned about the welfare of their crews; they simply refused to believe that they were placing them in jeopardy. The case of the *Milwaukee,* a sturdy car ferry that sank in a storm in 1929, illustrates how dangerous such confidence could be. Robert "Heavy Weather" McKay, a seasoned mariner, had brought his vessel across Lake Michigan in a foreboding storm, only to load up and head back out in the same storm. This time, his luck had run out. Was he as guilty as the captain of the *Griffon* about 250 years earlier? Or were both skippers, as so many in the years between the two sinkings, relying on bad weather experiences that Nature finally defeated? No one would ever know.

Vessels would continue to founder, ground in the shallows, collide with other boats, and catch fire; they sank under incredible circumstances, often after another circumstance weakened them. Sailors would tell you that they felt as safe on water as they did on land,

A wave breaks on the Chicago shoreline on November 10, 1913.

that their boats fared better than automobiles or even airplanes. Yet terrible accidents happen, whether you're delivering boxcars full of cargo on a carrier or Christmas trees on a schooner. If I've learned one thing working on many books about Great Lakes shipwrecks, it's that Nature usually wins, and that for all of the human efforts to sail in all conditions, human foibles can lead to disastrous results.

Lake Michigan waves crash on the Macatawa shoreline, 1913.

Phoenix (1847)

LIKE SO MANY PEOPLE settling into a new land, the Dutch arrived in the United States for a variety of reasons. They came to the country with hope: hope of escaping religious persecution, political strife, the famine facing their homeland; hope for economic prosperity, land ownership, a new and different life. They tended to follow the westward movement in the new land. Sometimes they accepted the recommendations of friends and loved ones who had preceded them. Sometimes they ignored warnings, chose their own paths.

It's difficult for most people today to imagine the hardship. Entire families, concerned or unconcerned about the difficulties they had heard of in letters from the new land, packed whatever they could carry, traveled to large port cities, and spent small fortunes on tickets to a new life. The trip across the Atlantic promised to be a trial and the new land unconventional. They would pay the exorbitant prices and stare down the risks.

There was great wealth to be accrued by the shipping companies transporting them, first across the ocean and second across the Great Lakes—to vast stretches of land largely unexplored or developed by white settlers, places that could present hostilities with little reward. The emigrants would be piled onto the boats. So many had no inkling of what ocean crossing involved: tiny airless spaces, seasickness and disease, poor food, no room to stretch out, weeks without seeing land. A small minority boarded ships and died before they reached the other side of the ocean; some who arrived questioned why they had come in the first place.

Crossing the ocean took a couple of months if the weather conditions were favorable. After that, many passengers transferred to vessels taking them to Buffalo, New York. Only then would they board the boat taking them to their final destinations. The vessel sailing on the Great Lakes needed to fulfill several requirements. It had to be small enough to fit through canals and locks but also large enough to carry a good number of passengers. The ability to move quickly was an additional bonus.

The *Phoenix,* a 140.6-foot wooden steamer completed and launched by the Pense & Alden company of Cleveland in 1845, was just such a vessel. She was constructed to carry both cargo and passengers on the Great Lakes, her typical journey moving from Buffalo to Chicago. She would also make the lengthy transatlantic voyage transporting European immigrants to the United States. Unlike the side paddlers and wind-driven schooners becoming prevalent on the lakes in the middle of the nineteenth century, the *Phoenix* was constructed with twin screws, a design considered radical at the time but soon adopted by other lake carrier builders.

The Buffalo–Chicago route was time-consuming, taking the *Phoenix* more than a week for her westward trip, often with stops along the way. Passengers onboard might go entire days without sighting land—and in cramped conditions, following an oceanic crossing, immigrants wanted nothing but the chance to settle in their new homes. Energetic children were hindered in their play by the constraints of the boat's dimensions and the many people onboard, all seeking breathing room. It meant little to the passengers in November 1847 that they were sailing across Lake Erie, up the

The *Phoenix*

St. Clair River, on to Lake Huron, through the Straits of Mackinac, and down Lake Michigan. The temperatures were falling and people looked only to their destinations.

Captain G. B. Sweet, a seasoned veteran of Great Lakes sailing, took the *Phoenix*'s helm on Thursday, November 11, his boat's final voyage of the season. Rough, unpredictable weather was possible this time of year, but the seas were placid as Sweet directed his vessel on the initial portion of the sailing across Lake Erie. The *Phoenix* was loaded to capacity with a cargo of coffee, molasses, chains, and hardware filling the hold; an estimated three hundred men, women, and children nailing down every square inch of passenger space; and a crew of twenty-five to guide the boat to its destination. Of the planned stops, Fairport, Ohio, was the first.

Smooth sailing didn't last long. Heavy wind from the north blew up formidable waves on Lake Erie, which was always a rough ride in stormy weather. The *Phoenix* held her course, even as seasick passengers, probably second-guessing their decision to sail on the lake rather than cross the land on foot or by wagon, suffered nausea. Then, when the *Phoenix* neared Fairport, Captain Sweet fell hard on the slippery deck and injured his knee. He was helped to his bunk, but rather than seek medical attention at one of the planned stops he decided to fight through agonizing pain and stay aboard the entire trip. With Sweet bedridden, First Mate H. Watts took over as acting captain.

The inclement weather continued. The *Phoenix*, although slowed by the heavy seas, sailed ahead. She made her scheduled stops in Cleveland and Detroit before heading up the St. Clair and, eventually, into Lake Huron. The small vessel took a beating but refused to surrender to the brutal punishment Nature administered. A steering-class window was lost to the waves, as were rail stanchions; water boarded the vessel. The battle took its toll on the boat's steam-powered engine and boilers, as greater-than-average supplies of wood were needed to keep the *Phoenix* plowing onward.

Hours melted into days. The weather relented, but only briefly, as the *Phoenix* made her westward turn toward the Straits of Mackinac. The boat had passed through the straits and entered upper Lake Michigan by the time misty weather visited again. Captain Sweet and First Mate Watts conferred in Sweet's cabin. The two decided to seek shelter on the lee side of Beaver Island and wait out the storm. There was no use in putting the *Phoenix* through more than she had encountered already.

The decision was a sound but difficult one. The crew might work on repairing damage to the *Phoenix* while the vessel was at anchor, but the wood supply was running dangerously low. Sweet and Watts estimated that there was barely enough wood left to reach Sheboygan, Wisconsin, the next scheduled stop. To make matters worse, gale-force winds forced captain and first mate to stay anchored an extra day, meaning they were even further behind schedule. Sweet weighed the positives and negatives and ordered the *Phoenix* back onto the water at the end of the next day.

The sailing remained horrible, prompting another decision: rather than sail to Sheboygan, the *Phoenix* would head to Manitowoc, which was roughly twenty miles north of her original destination. Sheboygan did not have a formal sheltered dock to protect the already battered boat, and the *Phoenix* risked more of a beating, perhaps a fatal one, if left to the elements. The *Phoenix* was scheduled to unload some of her cargo in Manitowoc on its return trip up the lake. Sweet reasoned the *Phoenix* could drop anchor in the eastern Wisconsin harbor, unload cargo and restore its wood supply, and then sail to Sheboygan. The possibility existed that the storm would blow itself out in the interim.

The boat inched its way into the Manitowoc harbor with marginal difficulty. Crewmen, aided by a handful of Dutch passengers, set out to load wood onto the *Phoenix*. It had grown dark, and the area was barely lit by the houses belonging to the town's scarce population. Passengers and off-duty crewmen walked to a lakeside bar for refreshment; women and children remained onboard. The first mate again consulted with his skipper, the two disagreeing about immediate plans. Watts favored remaining tied up overnight; Sweet, concerned that the boat was already very behind schedule, insisted on departure around 1:00 a.m. Watts reluctantly agreed. A number of Dutch passengers were destined for arrival, and new homes, in Sheboygan and were undoubtedly exhausted from their lengthy trip across the ocean and the Great Lakes. And there was the matter of the captain's getting medical treatment for his damaged knee.

The *Phoenix* departed on schedule, relieved of some of her cargo and replenished with six-foot cuts of wood. Predictably, most passengers were sleeping, tucked away for the final leg of their journey.

But something was wrong. Strange sounds came from the bowels of the boat. Most passengers who were awake, unaccustomed to the sounds created by the newfangled mechanizations of the screw-driven vessel, knew nothing. But one passenger, Clarence O'Connor, an Irishman with experience in working the boilers of a somewhat similar engine at a knitting factory, was troubled by what he heard. He dressed hurriedly in the dark and walked up to the deck. The lake had calmed noticeably, but O'Connor could detect no sign of land. The *Phoenix* was well out on the lake and possibly in trouble.

O'Connor rushed down to the engine room, where the first person he encountered was the *Phoenix*'s first fireman. He explained why he was there and his fears for the boat. The boiler, he said, was overheating and would explode. The fireman paid him no heed. The vessel's crew knew what they were doing. Who the hell was this Irishman, all worked up and acting as if he knew more than men experienced with these matters? O'Connor tried to push his way past the fireman. He was punched and pushed back for his efforts. O'Connor raced back to the deck and ran to the pilothouse. The second mate was in command. He was not convinced by O'Connor's explanation, either (he certainly wasn't taking orders from a passenger!), and he told O'Connor to leave. O'Connor retreated to his room, awakened his wife and daughter, and told them that they needed to dress and get to a lifeboat as quickly as possible.

It took little time for O'Connor's biggest fears to be realized. The intense heat from the boilers ignited the cordwood. The men in the engine room believed the fire could be easily contained and extinguished. They formed a small bucket brigade, pulling cold water from Lake Michigan and pouring it on the fire. The attempt failed. Thick smoke filled the air and flames burned everything wooden in their path.

Another much larger bucket brigade, consisting of crewmen and

A painting of the fire aboard the *Phoenix*

volunteer passengers, unsuccessfully fought the now out-of-control inferno. Flames escaped the windows, hatches, stairwells, and any other available opening. Passengers, now aware of the horrors visiting the *Phoenix,* flooded the spar deck, some screaming frantically in Dutch, children separated from their parents crying out for them. In the pilothouse, Second Mate Newall Merrick ordered a turn that would place the *Phoenix* on a course that might reach land—or at least shallow water—while there was still a chance of getting people off the boat.

This turned out to be an unrealistic expectation. The boat was totally without power. When the boilers shut down, the engine had gone as well. Dead in the water, the *Phoenix* could only drift. By all indications, she was not heading toward shore. She was engulfed in flames, some shooting as high as thirty feet into the air. Some passengers, desperate to escape the burning deck, scaled the boat's masts. Flames shot up the masts and the passengers were sent tumbling to the deck below. There was no escape, just the hopeless choice between perishing in the fire or jumping into the icy lake. Hundreds of people, including those on deck, screaming and crying and praying and shouting, so many in fiery bedclothes or choking on smoke, were doomed.

~

IT WAS TIME TO LAUNCH THE LIFEBOATS. Prior to the loss of the *Titanic* and the resulting focus on the requirement that each vessel contain enough lifeboats to accommodate every passenger, boats were pathetically inadequate at handling crew and passengers in a sinking. The *Phoenix* disaster occurred more than six decades before the loss of the *Titanic,* and in reality the *Phoenix* was in a position to lose a greater percentage of passengers and crew than the gigantic ocean liner. Accounts vary, but all agree that there were only two or three lifeboats on the *Phoenix.* Each boat carried twenty passengers.

Loading the lifeboats was a grim, herculean task. The chaos on deck, with all the shouting and running, increased when passengers saw the lifeboats being readied for loading and launching. People surged forward. Vessels had a protocol for lifeboat deployment, but people in fear for their lives and the lives of their children had no interest in process. Only a small minority of the passengers knew any English, and none of the *Phoenix* crew spoke fluent Dutch.

Out of the bedlam rose one heroic figure, a well-heeled businessman named David Blish. A married father, Blish hailed from Southport (now Kenosha), Wisconsin, and had accrued his wealth through the ownership of warehouses and docks in southeastern Wisconsin. He'd been in Holland on business, and during the lengthy trip across the Atlantic Ocean he had befriended many of the Dutch passengers. He spoke with the adults, played with the children, and even attempted to teach them rudimentary English.

After the fire broke out and it became clear that passengers had to abandon their berths, Blish focused on helping others, mainly children. He made his way belowdecks, crawling down the smoky corridors on all fours, calling out for children in Dutch. He fought off the choking and, eyes stinging from the smoke, would return, again and again, to the steerage levels. These efforts cost him dearly. He barely made it up on his final trip. Already suffering from smoke inhalation, he found the scene on the deck almost as bad as the one he'd just abandoned below.

A place on a lifeboat awaited him, due to his standing and his reserved cabin. He refused to board a boat, stating that he would take his chances on eventual rescue, that he was of more use helping others. This stance gained an ironic measure when crew members brought Captain Sweet to a lifeboat. Sweet vehemently refused to board the craft, saying that he wished to help passengers. Blish joined the fray, noting that the captain could barely stand, let alone be of any assistance. Safety rules required some officers and crewmen to be onboard lifeboats to guide operations and help with the rowing, whenever possible. Sweet was finally persuaded to be placed on board one of the boats.

Blish returned to the deck of the burning vessel. He was last seen jumping off the *Phoenix*, a child under each arm, into the freezing water. He never emerged from the lake's depths.

〰

ONE SHEBOYGAN RESIDENT, a judge named Morris, awakened from a deep sleep with a feeling that something was amiss. He had nothing to base the feeling on, but it was strong enough to get him to his feet. He padded over to a dresser and checked his watch for the time. It was shortly after two in the morning. The storm had subsided; all was quiet outside.

Not that it shouldn't have been. Sheboygan was a sleepy community, known more for its rustic farms than for industry. It had no harbor to receive supplies by boat; the best it could offer mariners was a modest pier jutting out over the lake. Even so, incoming vessels had to be small: the water was shallow, largely due to sediment deposited in the lake by the Sheboygan River.

When Judge Morris looked down from his home on a bluff overlooking the lake, he could see the shadowy silhouettes of two boats near the pier: a two-masted schooner and a small steamer. What he spotted out on the lake demanded more of his attention. Roughly four or five miles northeast of Sheboygan, an object glowed a bright orange-yellow. Morris squinted for a clearer look. He understood, to his utter amazement, that he was staring at a ship on fire.

He threw on some clothing and rushed to the pier, where he hoped to awaken the crews of the two vessels there. Mounting any significant rescue operation was bound to be a challenge. The wind and seas were now virtually calm, so the schooner would be hard-pressed to sail at much of a speed. The steamer would have as difficult a time leaving the pier: the boilers would require steam pressure to power the boat. This would take time, perhaps as much as an hour—an eternity to those souls struggling for survival on the burning vessel.

The men, tired from a short night's sleep, sprang into action. With no chance of either ship reaching the burning boat in time to save crew and passengers, the crew of the *Liberty,* the schooner, lowered their lifeboats and began rowing toward the disaster. Captain Tuttle of the *Delaware,* the steamer, ordered his engineer to create the steam necessary for departure. Tuttle had battled the storm just to reach Sheboygan. He could only pray that he could now quickly raise the steam necessary to sail in much calmer seas.

~

HOPE HAD BEEN ABANDONED on the *Phoenix* long before the dual rescue boats arrived. Burned to the waterline, the passenger liner was a vision of destruction. No one still on board survived. Victims littered Lake Michigan surrounding the destroyed vessel, the mortal remains of souls lost to drowning or hypothermia. Three survivors, partially immersed in bitter cold water, hung on and waited. Hope and despair, alternating with each minute, were addressed in silence.

Of the hundreds who had begun the trip on the *Phoenix,* only these three and those fortunate to find a place on the two lifeboats now approaching the Wisconsin shoreline remained. M. W. House, the chief engineer of the *Phoenix,* had fought the fire as long as he could before taking an ax to doors and walls and furniture—anything that could be tossed overboard and used as a flotation device. These actions met only marginal success. Some passengers drowned before they could reach the objects, while others climbed on whatever they found, only to succumb to hypothermia. House saved himself by using the ax to sever a rope attached to a fender, lowering himself to a door floating on the water and tying the door to the fender. It was a tenuous fix but offered hope if a rescue boat arrived in the near future. *Phoenix* clerk T. S. Donihue and a passenger from Milwaukee named Long survived by clinging to rudder chains on the boat's stern. Miraculously, these three stayed alive despite being partially submerged for more than three hours.

The *Phoenix*'s lifeboats staged dramas of their own. On one, a lost oar was replaced by a broom—an improvisation that worked until the lifeboat reached shore. The other lifeboat featured a "stowaway," a woman who was not allowed on the boat but swam over and silently hung on. No one knew that she was there. She made it to shore, somehow managing to fend off the effects of the frigid lake water.

For those arriving on the shore of Lake Michigan, the first order of business was building a fire—an irony that did not easily escape many of them. What little clothing they wore was saturated, and survivors were too cold to function. Those who could do so gathered driftwood; others scooped up anything that might be used as kindling. Soon enough, survivors huddled for warmth from the fire—what had been life-threatening only a couple of hours earlier.

Meanwhile, the sad chore at the site of the *Phoenix* wreckage continued. The crews of the two rescue boats located the three survivors in the water and brought them aboard the *Delaware,* where they were given dry clothing and food. A thorough search of the area found no other survivors.

Captain Tuttle decided to tow the *Phoenix* to Sheboygan. The burned-out boat was still valuable: the engine and boilers might still be salvageable, and there was no telling what other valuables were in the wreckage. The safe, of course, stored the most precious of passenger holdings. The Dutch immigrants had brought their money and valuables to the new land to establish what they planned to be new lives. Some had sewn valuables into their clothing or stowed them in money belts, all left behind in the frantic efforts to abandon the burning boat. These items might be found and claimed by salvagers or ordinary citizens willing and able to explore the wreckage.

Tuttle ordered his crew to attach a towing line to the *Phoenix*'s stern. The boat was still burning and might not survive the trip to dry land, but the short journey would be well worth the effort. The storm had blown itself out and the water surface was smooth. The gruesome task of collecting bodies would begin soon enough.

The *Phoenix* wound up grounding in eight feet of water near Sheboygan's pier. Shortly after sunrise, townspeople began to gather for a look at the wreckage. An untold story lay in the ruins.

EIGHT *PHOENIX* CREWMEN made it safely to shore, including the boat's captain, first mate, and chief engineer. They had witnessed the tragedy almost from the beginning. In the aftermath, they discussed and debated how the fire started, why it had not been extinguished or contained, why it spread so quickly; they wondered why the boilers had died with the boat so far from shore. House accurately described the confrontation between Clarence O'Connor and the boat's second engineer. O'Connor had been justified in his concerns, and he and his family had survived while the second engineer had not.

There were bound to be formal inquiries demanding answers, as they had none. They were irked by rumors, circulating almost from the beginning, that some of the crewmen visiting the bar in Manitowoc had been too intoxicated to perform their duties under the duress of an emergency.

They could only guess at how many people had perished. Precise passenger and crew manifests were not a high priority in the mid-1800s, and any records burned with the vessel. The final count: forty-three survivors, one hundred to two hundred dead. No one would ever know the exact totals.

The survivors were picked up and transported by wagon to Sheboygan, where they were given complimentary rooms at a hotel. Transportation was arranged for anyone traveling outside Sheboygan.

Scavengers converged on the *Phoenix* in the days following the disaster. Talk of riches in the vessel's safe, treasure chests, money belts, and clothing yielded only small rewards. The safe was never recovered, despite numerous diving efforts, and no treasure chests emerged. Sheboygan's coroner, James Berry, visited the wreck site and found a couple of money belts containing gold. Gold coins were found scattered on the lake floor near the wreckage. Word circulated that Berry used the money to purchase prize Holstein cattle, supposedly the first in a state that would become famous for its dairy.

Captain Tuttle found headlines in unwanted ways. After sailing from Sheboygan, he guided his vessel north to the Manitowoc area where the *Phoenix* had caught fire. Bodies were scattered throughout the area. Despite pleas from his crewmen, Tuttle refused to recover the bodies. He was already behind schedule. Besides, he told complainants, the remains would be picked up by local fishermen. When word of this got out, Tuttle's reputation took a severe hit.

The twenty-four Dutch survivors, destitute in a new land, were taken in by other Dutch settlers in Sheboygan. They found jobs, rebuilt their lives, and, in most cases, stayed in the area, bought land, and farmed. Their lost families, friends, and fellow travelers were never forgotten, their stories sewn into the fabric of family history. The last of the survivors died in 1918.

THE DEMISE OF THE *PHOENIX* remains one of the greatest tragedies in Great Lakes maritime history. An extended period of mourning followed the news of the vessel's loss, and the disaster was never far from the minds of the survivors. The passing years, however, had a way of fading memories. With each anniversary, the *Phoenix* story, although told and retold, received less emphasis than in preceding years.

A waterlogged Bible, published in 1673, floated ashore. It was eventually placed in the Wisconsin Maritime Museum. A scammer claiming to possess a load of burned wooden shoes that supposedly washed ashore was run out of Sheboygan when he tried to sell them to locals.

The *Phoenix* wreckage suffered an ignominious fate. A winter storm dismantled her bow just months after the loss. The following spring, a farmer chopped up the bow for firewood. A few months later, the engine and boilers were removed from what remained of the wreck, which slowly disappeared from sight.

Shipwreck hunters searched for her remains, with no luck, until 2022. A Dutch documentary crew and a shipwreck hunter found the *Phoenix*'s smokestack in 85 feet of water, not far from where she had come to rest before being towed to Sheboygan.

On May 29, 1848, six months after the burning of the *Phoenix*, Wisconsin was admitted to the union as its thirtieth state.

The *Phoenix* tragedy marker in Sheboygan, Wisconsin

Niagara (1856)

FEW SAILING POSSIBILITIES struck fear into the minds of Great Lakes captains more than the threat of a fire or boiler explosion. The 1847 loss of the *Phoenix* on Lake Michigan had been a sobering reminder that great loss of human life could occur even a short distance from shore—and it was by no means an exception. As Great Lakes historian Mark L. Thompson points out in his book *Graveyard of the Lakes,* in 1823 alone "boiler explosions destroyed 14 percent of all steamboats operating in the United States." There weren't as many steamers making their way around the lakes at that early date, but the percentage is still significant.

The figure might be attributed to the gulf between old and new shipbuilding practices. Vessels were still constructed of wood, and this would continue for the rest of the nineteenth century. Wind moved the boats, but that changed in 1817 when two boats, the Canadian *Frontenac* and the American *Ontario,* became the first boiler-driven vessels on the Great Lakes. This created a formula for disaster: the wooden vessels, covered in highly flammable paint and lacquer, were powered by engines dependent on fires stoked in boilers reaching high temperatures. Safe boilers were still in the development stage when the early steamers sailed.

Controversy was inevitable, as was a clash between business and safety. With the Midwest expanding enormously as immigrants came to the United States, the need for transportation was more important than ever. And it was not only the influx of population: there was an urgent demand for the supplies required for building cities.

New safety standards were issued, including requirements for lifeboats, firefighting equipment, regular boiler inspections, trained engineers overseeing boilers and engines, and other precautions. Many of the new laws were enacted by Congress in 1838, but in essence they meant very little. Shipping companies always resisted change, especially change involving notable financial investment. Owners and officials of those companies had the clout to appear concerned about safety and the muscle to see that any new laws, while paying lip service to proper measures, also had little to enforce them.

Tremendous losses with high fatalities continued. On August 9, 1841, the highly regarded *Erie,* carrying more than two hundred passengers on Lake Erie, caught fire and sank during a desperate attempt to reach shore, with only twenty-seven survivors. The four-year-old, 176-foot side paddler had received fresh coats of paint and varnish only days before the fire. Six years later, the *Phoenix* was destroyed. In 1850, the three-year-old *G. P. Griffith* caught fire; more than 240 of its 326 passengers, mostly immigrants traveling to their new homes, perished.

The mind-numbing list of losses from fires and boiler explosions expanded with the increasing number of steamers transporting passengers around the lakes. More safety measures were introduced in the industry, with little effect. The steamship had yet to be perfected. As Thompson reports in his book, there was almost a sense of resignation to the dangers of steam-driven vessels dating back to 1824, when Treasury Secretary William Crawford declared that "it seems as though more steam boilers blew than ran."

Serious problems notwithstanding, passenger boats served a crucial function in carrying immigrants to ports throughout the Great Lakes. The vessels varied greatly in quality and service. Some stacked passengers like cordwood with little regard for comfort; it was all about quantity. A ticket was less expensive than one might purchase on other lines, but the more lower-paying passengers who could pile onboard, the better the earnings. The more luxurious boats charged more and reaped their own benefits.

The luxury liners were known as palace steamers. They boasted of large, well-appointed cabins with deep carpeting and fine furniture. They had ballrooms with stained-glass domes. They offered exceptional food. The beautiful side paddlers drew raves from their passengers and from casual observers. Oddly enough, they got their start as troop transports during the War of 1812.

The *Niagara,* owned by the Collingwood Line, ranked among the finest palace steamers in the business. At 230 feet, she was also one of the longest at the time of her launching. Constructed by Bidwell & Banta in 1845, the *Niagara* carried freight as well as passengers, many of whom were immigrants settling in the Midwest. People would line the shores and watch when she was in the area.

Nine years had passed since the loss of the *Phoenix,* but the story of that ill-fated boat was undoubtedly known by the crew of the *Niagara.* The same could be said about the hazards and possibilities of fire in general. The *Niagara* had supposedly been built with a fireproof engine room, eliminating the main source of many boat burnings; still, the vessel's officers were less than vigilant about taking on flammable cargo banned by the industry.

When the *Niagara* pulled out of the Collingwood, Ontario, harbor on September 24, 1856, bound for Chicago with a full load of passengers and cargo, there was no reason to suspect that the run would be anything outside the ordinary. Roughly three hundred passengers (one hundred in first class, two hundred in steerage) relaxed as the boat made her way down Lake Michigan at a good, steady clip. The weather, though cold, was favorable—certainly nothing the many immigrants on board weren't accustomed to in their native lands.

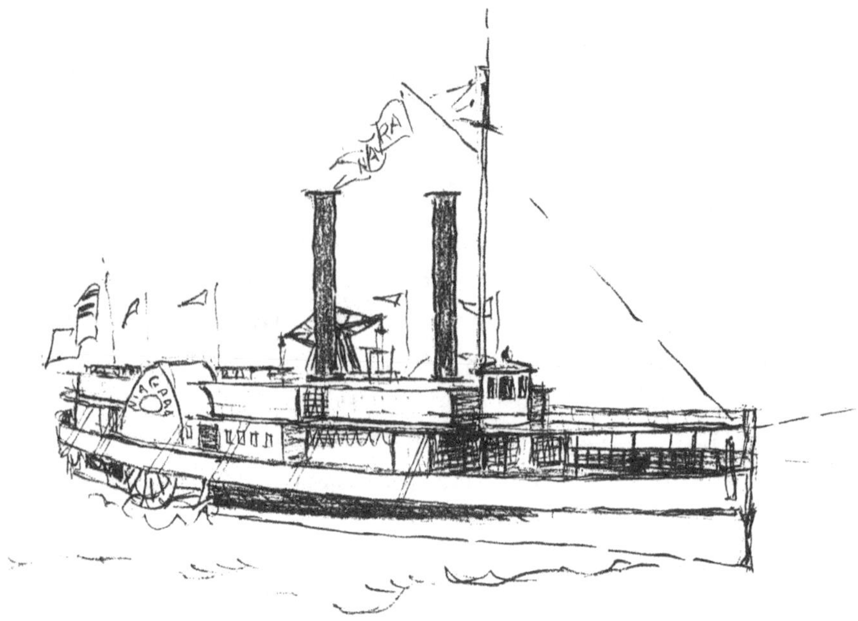

The *Niagara*

The *Niagara* dropped off more than half of her passengers in Sheboygan, Wisconsin, the city that had experienced the *Phoenix* horrors less than a decade earlier. The next scheduled stop was Port Washington, Wisconsin, with Chicago only thirteen to fifteen hours away.

Somewhere between Sheboygan and Port Washington, black smoke was spotted coming from the area of the engine room. It was late afternoon. Captain Fred S. Miller, off duty and resting in his quarters, was awakened by shouts of *"Fire!"* and charged to the deck to see what the commotion was all about. Pandemonium reigned. Thick smoke poured from the bowels of the vessel. Terrified passengers rushed everywhere. Flames were not immediately noticeable, but that would change quickly.

Miller ordered his men to use the fire hose and then hurried to the pilothouse to further instruct the crew. Incredibly, not a single life jacket was on board the vessel, which only magnified the panic. Miller told his wheelsman to direct the *Niagara* toward shore—an order that may have worsened the boat's peril, as the breeze caused by the *Niagara*'s hastening through the water only served to fan the flames now engulfing the vessel.

By this point, passengers were out of control. Children separated from their parents cried for help. Passengers began hurling anything that might float overboard. Some tried to get away by jumping into the water or lowering themselves on the boat's ropes. Others drowned instantly when tossing their children and following them into the freezing water without life jackets. The only hope appeared to be finding something that floated and hanging on until a rescue vessel arrived.

The few lifeboats proved to be useless. In the bedlam, more passengers headed to the boats than possibly could have been accommodated, and overloaded lifeboats crashed into the lake, overturned, and splintered. John Macy, a wealthy former congressman from Wisconsin, offered $100,000 to anyone who could save him. When he found no takers, Macy tried to leap into a lifeboat overloaded with women and children. The boat crashed into the lake, drowning everyone, including Macy. It was later estimated that more victims died from drowning than from fire. As the fire raged, explosions belowdecks could be heard, possibly the result of flammable cargo igniting. There is no question that this contributed to the fire already out of control.

Tales of heroism rose from the ashes. A two-year-old child, separated from his parents and standing near Macy, was swooped up by another passenger, who braved the fire that was nearly splitting the *Niagara* into two separate sections. The man (whose name was never learned) leapt into the water with the boy in his arms. He found some floating wood, climbed aboard, and paddled toward shore with the boy. The next morning, the two were found, exhausted but safe and alive, on the beach near Port Washington.

Fortunately for those still trapped on the *Niagara,* the boat was within eyesight of land when she lost power. Two steamers, the *Illinois* and the *Traveler,* saw the vessel's plight and, joined by a couple of schooners, lit out to rescue any survivors. They lowered their lifeboats and people were pulled out of the lake. Captain Miller and his crewmen were saved. On a historical note, Captain Frederick Pabst of the *Traveler* went on to found the longstanding brewery of the same name.

MYSTERY CLOUDED the *Niagara* story. Since exact passenger lists were rare at the time, there were only estimates of the number of people lost; some estimates exceeded one hundred, but the generally accepted figure rounded out to sixty. There was never a final determination of what exactly started the fire. Captain Miller insisted that it had not started in the engine room and must have begun in the cargo hold, but there was even talk that arson might have been involved. A note, believed to be a hoax, was found by the *Niagara*'s steward in his room:

> Look out. Save yourself. The boat will be burned tonight. Everything is in Readiness.

Whatever the case, the burned-out hulk of one of the lake's most luxurious liners sank in about fifty-five feet of water not far from Port Washington and what is now Belgium, Wisconsin. Its remains have disintegrated, leaving only questions and the certainty that the *Niagara* would be one of the worst maritime disasters in Great Lakes history.

Lady Elgin
(1860)

When one thinks about deadly shipwrecks on the Great Lakes, it's easy to imagine aging vessels meeting their demise in storms, collisions, fires, or other accidents. These boats served admirably but, because of new methods and materials used in shipbuilding, fell behind and were more susceptible to fatal problems.

This was not the case in the loss of the *Phoenix*, still rather new at the time of her burning, or of the *Lady Elgin*, which was less than ten years old when she became a tragedy of unspeakable proportion in 1860. She had a strange background, but nothing that might suggest to the best-known soothsayer that she would reach her end the way she did.

Constructed in Buffalo by the Bidwell & Banta shipbuilding company and launched in November 1831, the *Lady Elgin* was named after the wife of the former governor-general of Canada. Her existence was not without color: Grand Trunk Railway, her Canadian owner, viewed her value as much for hauling freight as for transporting passengers, with the main body of her work involving sailing on Lake Erie. The *Lady Elgin* expanded her service five years later, when she was sold to A. T. Spencer & Company, a business that wanted her for passenger and freight runs between Chicago and Lake Superior. This made the beautiful double-decker built of white oak a welcome, familiar presence on Lakes Michigan and Superior. Spencer & Company sold her in early 1860 to Gurdon Hubbard, a Chicagoan who maintained her established route.

These sales may or may not have been the result of the *Lady Elgin*'s checkered (and costly) sailing history. The Midwest was growing, with cities such Chicago, Detroit, Milwaukee, and Minneapolis ballooning into much larger municipalities and assuring a constant need for building supplies and passage for immigrants looking for new, and possibly lucrative, starts. This resulted in a full schedule for the *Lady Elgin* and other boats, which meant more risks for accidents and mishaps.

The *Lady Elgin* had been around for only three years when she hit a large rock and sank near Manitowoc, Wisconsin. She caught fire in 1857, and though she survived this, too, she was notably damaged. A year later, she grounded at Copper Harbor, Michigan. Then, the following shipping season, she grounded again at Au Sable Point on Lake Superior. Mechanical problems led to her being towed on two other occasions. She was always repaired to look as good as new, but to superstitious sailors, she seemed cursed.

POLITICS PLAYED A CENTRAL ROLE among the passengers gathering dockside to board the *Lady Elgin* on the evening of Thursday, September 6, 1860. Slavery dominated conversations in the presidential election year. Abraham Lincoln, a Republican, was running on a strict abolitionist platform: he would abolish slavery nationally and federally. One of his most outspoken opponents, Stephen A. Douglas, an Illinois senator who had defeated Lincoln for his office two years earlier following a series of historic debates, also abhorred slavery, but he strongly felt that the issue was best settled by individual states rather than the federal government. He was scheduled to deliver a speech in Chicago on September 7. As part of a weekend series of events, it promised to be well attended.

The *Lady Elgin*

The people of Wisconsin had their own strong, conflicting opinions in the slavery debate. Governor Alexander Randall, a Republican, made no secret of his burning hatred of slavery. He spoke often—and forcefully—against it, going so far as to declare that Wisconsin, though a state only for a dozen years, would secede from the union if the federal government didn't abolish slavery. Further, he polled the state for support and opposition. Whenever possible, he found ways to dispose of those who disagreed with him, especially within the state militia. Randall had his adjutant general check different groups for support; opposing units quickly found themselves without money, armaments, or any real power.

Milwaukee had four militia units opposing Randall, and in one, the Union Guard of the city's Third Ward, Randall discovered that he had a worthy adversary. Captain Garrett Barry was old school, a West Pointer with an impressive military background. He had fought with Zachary Taylor in the Mexican–American War. He eventually married and moved to Milwaukee, his interest in politics rewarded in elections to the positions of treasurer of Milwaukee County and superintendent of the post office. Like Douglas, he vehemently opposed slavery but believed it was a state, rather than a national, issue to resolve. Randall's threat of secession, he believed, was treasonous. He didn't shy away from making his position public.

Randall responded by trying to disband the Union Guard. He disarmed it and took away state interest and support. This didn't work. Infuriated, Barry kept his mostly Irish membership together. The events on September 7 in Chicago would constitute a Union Guard rallying cry. His militia would attend parades, marches, Douglas's speech, and social gatherings in full uniform, and they would use these occasions to raise funds for new weapons he had obtained in St. Louis and for future operating expenses.

Barry knew of a vessel, the *Lady Elgin*, downbound from Lake Superior to Chicago and scheduled for a stop in Milwaukee. She would be departing from the Milwaukee dock at 7:00 p.m. on September 6 and arrive in the Windy City in plenty of time for the Union Guard to attend all planned events. Barry reserved a large block of one-dollar round-trip tickets. They sold at a brisk pace, more than three hundred in all, assuring the Union Guard of a strong showing in Chicago.

~

THE *LADY ELGIN*'S Milwaukee-to-Chicago run was popular. It was relatively brief and pleasant, promising businessmen and politicians the opportunity to engage in their various interests in short order. The boat offered every amenity: Live bands and orchestras entertained passengers dancing in spacious halls under beautiful chandeliers. Men could chomp cigars in elegant salons. The decks were large enough to accommodate one hundred guests willing to sleep under the stars. The staterooms seemed to have been designed for royalty. The food was memorable. There was little wonder why the *Lady Elgin* was nicknamed "Queen of the Lakes."

Captain Jack Wilson took justifiable pride in his vessel. A veteran of nearly a quarter century of sailing, Wilson had commanded other boats prior to taking over the *Lady Elgin* in 1857—quite possibly, notes Benjamin J. Shelak in *Shipwrecks of Lake Michigan*, "as a gesture of appreciation for his assistance in building the [Soo] Canal." The

Canal, completed in 1855, finally offered passage between the lower lakes and Lake Superior, opening up commercial shipping possibilities previously unheard of. A gregarious man, Wilson enjoyed mixing and trading stories with the *Lady Elgin* passengers. He was well respected by his crewmen, who seemed to enjoy the straightforward Lake Superior–Chicago run.

This early September 1860 trip, however, was running well behind schedule. The Milwaukee-to-Chicago passengers were growing impatient as the hours passed and the *Lady Elgin* had yet to arrive. Had the boat run into some kind of trouble? Was she undergoing repairs? What could be taking so long? The Union Guard, dockside early for the boat's arrival, grew edgy, though many left to visit a local tavern for refreshment.

The delay was the result of natural causes: Captain Wilson had checked down when his vessel ran into heavy fog. The *Lady Elgin* reached the Milwaukee dock nearly five hours late, unloaded cargo and passengers, and had to stand by while taking on the large number of new passengers. The boat was technically overloaded (she was supposed to take a maximum of only three hundred passengers), but Wilson could think of no reason to hold down his passenger manifest for such a short trip. He assured those boarding that they would be in Chicago by early Friday morning. By midnight, Captain Wilson was guiding his charge out of the harbor and onto Lake Michigan.

The *Lady Elgin*'s sail to Chicago was uneventful. She arrived just before sunrise, to the relief of everyone with plans for that morning. The members of the Union Guard were pleased that all the day's plans were still in place.

And so it went. Aside from hearing speeches, attending parades and assemblies, and other adventures, the Union Guard had plenty of time for sightseeing and fulfilling fundraising obligations. These activities made for an exhausting day, but the large group was prepared to party all the way home. They had the rest of the weekend to catch up on sleep.

They were unhappy to learn that there might be alterations to their plans: again, the weather was not cooperating. The dark, cloudy sky gave an ominous threat of a storm. The barometer fell. Captain Wilson, well versed in the temperamental ways of autumn storms on the lakes, did not like what he was seeing. He took great pride in the fact that, in an era of accidents, he had never lost a crewman or passenger on one of the boats he was commanding, and he was not inclined to take risks now. He considered keeping his boat docked for the night, letting the storm pass, and starting for Milwaukee in the morning, but this idea left his passengers, especially the Union Guard, sour. Wilson relented. Shortly after 11:30 p.m., he eased the *Lady Elgin* up the Chicago River and onto Lake Michigan.

The storm turned out to be everything Wilson feared it might be. Heavy rain pounded the *Lady Elgin* mercilessly. High winds stirred up the water. The boat pitched and rolled, the wheelsman struggling to keep her on her northern course in gargantuan waves. The passengers didn't seem to notice. They reveled like it was New Year's Eve, dancing to a German brass band and drinking heartily.

Elsewhere on the lake, the *Augusta,* a 128-foot two-masted schooner overloaded with lumber in her cargo hold and on her deck, waged her own war against the maelstrom. Her voyage from Port Huron, Michigan, to Chicago was long and, given the roiling seas, treacherous. The wind tore at her sails, threatening to capsize the boat. Captain Darius Nelson Malott, who had been asleep in his bunk shortly after midnight, consulted with his second mate, George Budge, about sending crewmen to face the perils of trying to take down the *Augusta*'s sails in a storm that was tossing his boat around like a toy.

Malott, a farm boy in Canada in his youth, had unbelievable experiences with water—the type that Herman Melville might have spun into a novel. He took a job as a deckhand on his eighteenth birthday and sailed on the lakes for a couple of years before ultimately signing on with the *Gold Hunter,* a saltwater ship hauling lumber to London. A *Gold Hunter* journey almost ended his life. Caught in a fierce Atlantic storm, the ship lost her mast and rolled over onto her side. The captain and first mate were washed overboard and drowned. Helpless, the survivors drifted on the ocean for forty-five days. The provisions ran out before there was any sign of rescue. The starving men decided to eat one of the two single sailors who had no familial responsibilities—one of whom was Malott. He and the other man drew straws, and Malott was temporarily spared. To his great fortune, a mail ship came by and he was saved when it took the lost sailors to London.

Malott's travails weren't behind him. He signed on with another salty, the *Edward Hyman,* and eventually set sail on a journey east that would take him first to Hong Kong and then across the Pacific Ocean. While on the trip, he was promoted to second mate. Two days later, the ship caught fire and two lifeboats holding the crew were launched. The captain commanded one boat, Malott the other. The two were separated, and the captain's lifeboat was never seen again. Malott and his group, fortified by only minimal provisions, drifted until they reached shore in Peru. Malott had been spared again. He was twenty-one years old.

He moved back to Canada and married. In 1858, he was hired as master of the *Caroline,* another salty. This time, matters were less hazardous. Malott took a load of cargo to London, oversaw the sale of his vessel, and came back to the United States. A year later, he was awarded command of a five-year-old schooner, the *Augusta.* He would be working the lucrative lumber trade.

Load sizes determined profits, and for his Port Huron–Chicago run Malott loaded as much lumber as his vessel could take—an acceptable move in good weather, but a questionable one if the weather turned ugly. The *Augusta* was no match for the storm in September 1860. Lumber shifted, and she tipped almost onto her side. Once again, Malott feared for his life—and on his birthday, no less. His boat was out of control, taking on water, careening into oblivion at a high rate of speed. Gale-force winds caught his sails and looked to be capsizing the vessel.

Malott ordered Budge to take some crew members outside and lower the *Augusta*'s sails. It was their only chance. The crew were laboring at their job when Budge saw what he couldn't believe: dead ahead, no more than a quarter mile away, were the running lights of a large vessel. Budge shouted for his captain, but it was too late. A collision was unavoidable.

~

JOHN JARVIS, a passenger on the *Lady Elgin,* was standing on deck when he saw the *Augusta* heading toward the *Lady Elgin,* about to hit her broadside. "I was standing at the middle gangway," he said later. "I saw the schooner about two minutes before the collision. She appeared to be coming toward us at an angle of forty-five degrees, about half a minute before she struck."

John Verce, first mate of the *Augusta,* was out in the storm trying to bring down the sails when he heard the captain call "Hard up!" But the *Augusta* didn't respond to the attempts to steer her free of a collision. "I did not notice any change in the vessel's course," Verce recalled later. "The *Augusta* steers wild—that is, will not answer her helm readily. It was not two minutes at the outside after the order till the vessel struck."

The *Augusta* slammed into the *Lady Elgin*'s port side, just off the side paddle, tearing a huge gash into the larger vessel. The hole—big enough "you could drive a team of horses through it"—extended beneath the waterline. Both boats shuddered violently from the collision. The *Augusta* remained embedded, at a right angle, in the *Lady Elgin*'s side with the *Lady Elgin* dragging her along as she was driven by the waves, the two enjoined for a short distance. Then the wind and waves separated then. The *Lady Elgin* sailed on into the night.

Captain Wilson recognized that his vessel might be in mortal trouble. Water gushed into the boat as soon as the two vessels separated. The engine room was flooding. Wilson instinctively understood that, barring a miracle, it was only a matter of time before the *Lady Elgin* was at the bottom of the lake. And what of the other boat? Why hadn't she stuck around to offer assistance?

As Malott would later state, it was mainly a matter of misunderstanding. Watching the *Lady Elgin* sail on, Malott reasoned that his vessel must have struck a glancing blow—or at least one above the waterline. The steamship was continuing on as if nothing had happened. And Malott had worries of his own: the collision had pushed in the front of the *Augusta,* though Malott was relieved to learn, after sending crewmen to survey the damage, that his boat was minimally

A sketch of the *Lady Elgin* colliding with the *Augusta*

wounded. His mission now was to race to Chicago and hope to outlast the storm. He was still facing blinding rain and huge waves.

Captain Wilson initially attempted to find a way to keep his vessel afloat, either by plugging the tear or by leaning the *Lady Elgin* so the tear would be out of the water. He had no certainty about the extent of the damage. He ordered First Mate George Davis to gather crewmen on a small boat, lower it to the side of the boat, check the damage, and ascertain whether the hole could be blocked. This would have been a tall order in placid weather; in the storm, the small boat was impossible to negotiate in the waves. Lightning provided their only illumination in the darkness. Frightened passengers milled about, trying to figure out what had happened and what kind of danger they were in. Lowering a small boat was not an encouraging sign to those wondering about the status of the vessel.

In the pilothouse, Wilson proposed other possible solutions to the rapidly deteriorating situation. He had a watchman clang the boat's bell to alert all passengers of the vessel's perils. The *Lady Elgin* had brought livestock on board, fifty cattle, for the return trip; they were penned on deck. Wilson ordered crew members to push them overboard, if possible—anything to lighten the load. He ordered his wheelsman to steer the boat toward shore, though that, too, might have been out of desperation. He estimated that, at time of impact, the *Lady Elgin* was roughly ten miles from shore. The ship was now listing badly and sinking into the lake. He asked passengers to assist in distributing life preservers.

Wilson had another small boat launched to help the others with plugging the hole with mattresses, but, like the first boat, it was defeated by the heavy waves and tossed away from the *Lady Elgin.* A boiler broke through the bottom of the boat, sending in even more water. The *Lady Elgin* began breaking apart. Her hurricane deck, chopped at by crew members trying to create something to hold many passengers at once, was torn loose by waves and hurled in one large piece into the lake.

The end came quickly: survivors estimated that only twenty minutes passed between impact and sinking. Strangely enough, the passengers remained relatively calm, despite the severity of their plight. There was the predictable shouting and screaming, but survivors later spoke of the orderly way in which people behaved.

H. G. Clavyl, clerk of the *Lady Elgin*, recalled the calm a short time after the collision. He roamed the boat and was amazed by what he saw. "I passed through the cabins," he stated. "The ladies were pale, but silent. There was not a cry or shriek—no sound but the rush of the steam and the surge of the heavy sea. Whether they were not fully aware of the danger, or whether their appalling situation made them speechless, I cannot tell."

Much of this might be attributed to Captain Wilson, who abandoned the pilothouse and moved among the passengers. Mustering as much serenity as he could, he reassured them that help was on the way, that if they wound up in the water they should cling to anything that floated, and that the waves would propel them toward shore. In the final minutes of the *Lady Elgin*, anything floatable was thrown into the water. The hurricane deck, still within reach, became an oversized raft, and the band's drummer ended up riding his bass drum to safety.

The boat split into three pieces. The bow section stayed afloat; the two aft portions sank immediately. Most of the passengers, maybe as

A sketch of the *Lady Elgin* shipwreck

many as three hundred, were in the water. Those who couldn't swim drowned. Most hung on to boards, trunks, pieces of doors or deck—anything they could locate and grasp. More than fifty climbed aboard the floating deck. The waves moved them swiftly through the cold water. Rain fell. Lightning continued to flash. In the dark, wee hours of the morning, hope was all the survivors could bargain for.

CAPTAIN MALOTT AND HIS CREW did not witness the *Lady Elgin*'s sinking as they sailed in the opposite direction down northern Illinois toward Chicago. They had no idea of the identity of the boat they had hit and, in fact, assumed that it had sailed on without a problem. It was only the next morning, after the badly battered *Augusta* docked in the Windy City, that Malott learned which boat he had collided with. By then, most of northern Illinois was aware of the sinking.

For those awaiting rescue, the night had been horrific, but nowhere near as bad as the daybreak hours the next morning. As Captain Wilson had suggested, the storm rapidly pushed the survivors toward shore. He had been lucky enough to find a place on the floating deck and maintained the air of authority useful in reassuring the large group hanging on for dear life. His calm demeanor steadied those who might have otherwise given up. There was no reason to surrender, he told those assembled on the raft; they would soon be safe again.

He was at least partly correct in his assumption. The first of the boats lowered when the *Lady Elgin* was sinking, the one commanded by First Mate George Davis, found its way to shore at daybreak. It arrived just north of Winnetka, a sleepy little village not far from Evanston, the college town where Northwestern University was located. Davis and the others scaled the bluffs near the lake and moved from door to door, telling the startled inhabitants of the sinking. The survivors—and there might be many—were going to need as much help as they could get. The surf near shore, though somewhat shallow, was rocky and lethal. People might survive the lake's depths only to perish with safety within sight. Volunteers responded. They brought rope, dry clothing, and food, and they made their way to the beach.

The word spread to Evanston, and people there took action as well. Edward W. Spencer, a divinity student at Garrett Biblical Institute, heard of the need and threw on some clothes. A slightly built man, Spencer was an excellent swimmer. He believed he could assist, and he would become one of the day's biggest heroes.

Edward W. Spencer

A former Winnetka alderman, Artemas Carter, vividly recalled the scene on the beach as the rescue unfolded. "We could see small objects to the northeast floating or rather pitching and tossing on the mad waves," he said. "As they came nearer, we saw that they were fragments of the steamer, and freighted with human beings—some with one person, some with two, three, or a dozen persons. People began to rally on the shore with ropes."

A sketch of Edward Spencer and crew during rescue efforts

For the shipwreck castaways, the scene must have seemed surreal. They had survived the sinking of a ship and a lengthy passage over storm-ravaged waters only to face death within clear view of so many assembled to save them. Their journey had seen their numbers diminish: waves had separated survivors from whatever they were using to stay afloat, and the violent seas had also smashed wooden objects to splinters. A heavy undertow had pulled people down. But the most serious damage was done within one hundred feet of shore.

The hurricane deck offered evidence of the storm's effects. Big and strong at the time of the *Lady Elgin*'s sinking, the deck had been whittled down to almost nothing. About one-third of its original passengers, including Captain Wilson, remained when the makeshift raft approached the surf near shore. Two women slipped overboard, and Wilson went in after them. None of them returned to the surface.

The onshore rescuers often found their efforts rewarded. They heaved ropes out to those caught in the surf. When the nearly drowned survivors latched on to the ropes, parties of people onshore would pull them in. The rally, as Artemas Carter labeled it, could be credited with some success.

This was especially noteworthy in the case of Edward Spencer. When he arrived on the beach, he tied a long rope around his waist and, with fellow students anchoring him, fought his way to the surf. Over and over—seventeen times in all—he returned to the site and rescued exhausted survivors. On one occasion, he was struck by a piece of wood, which opened a nasty cut on his head. On his final trip out, Spencer saved two people at one time, a husband and wife who almost succumbed to the waves.

Spencer's heroics took their toll. He was so badly weakened that he collapsed after his final struggle to the scene. Delirious, he kept asking those on shore, "Did I do my best? Did I do my best?" He never fully recovered. According to the *Chicago Tribune,* he would live the rest of his life a "semi-invalid." He dropped out of school, moved to California, married, had children, and lived a quiet existence. A plaque honoring him and his efforts was erected at Northwestern University in 1898.

~

A TRAGEDY OF THE SCALE of the *Lady Elgin* demanded a full-scale investigation. It remains the greatest loss on open water in Great Lakes history. Final victim numbers will never be known; there is no passenger manifest, and even if there were, stowaways (many on board to join festivities, still there when the *Lady Elgin* pulled out of Chicago) were lost. The estimates vary, but there is no questioning that more than two hundred lost their lives, with some figures creeping up to three hundred. Approximately one hundred survived.

Word of the shipwreck circulated in a hurry throughout the northern coast of Illinois and the Great Lakes region and continued long after the surviving Union Guard members found their way back to Milwaukee. The ship loss had decimated the guard, and the victims' families took the train to Winnetka for the grisly task of identifying loved ones. When word leaked about the *Augusta*'s role in the accident, angry members of the Union Guard swore revenge.

The Winnetka train depot served as a temporary morgue until victims could be identified and returned to their homes. In far too many cases, victims would be identified, but the money in their pockets had been removed by scavengers. Thieves had combed the area for money and other valuables as soon as the bodies hit land. Captain Wilson's remains washed ashore three days later after being carried by waves to Michigan City, Indiana. Other corpses, as well as wreckage from the *Lady Elgin,* washed onto land throughout the northern coast of Illinois. Well-known passengers included William Farnsworth, a founding father of Sheboygan, Wisconsin, and Herbert Ingram, a London newspaper owner and member of Parliament. It was said that one thousand children were orphaned in the accident, though that figure seems exaggerated today.

The *Augusta*'s part in the tragedy was given a thorough examination. Authorities grilled Captain Malott about the factors leading to the collision, particularly whether the *Augusta* was visible to those in the *Lady Elgin*'s pilothouse in the minutes before impact. There was some question about whether the schooner was using its signal light, but Malott and his crew insisted that the light, in working order, was in use—a claim backed by the *Lady Elgin*'s second mate. Malott admitted that his vessel was out of control, caught in a trough of wild seas virtually impossible to navigate.

The twelve members of a hastily called Cook County coroner's hearing listened intently as crews of the vessels recalled what happened before, during, and after the collision. They accepted the testimony that both boats were using their lighting equipment, that the *Augusta* could be seen at least a quarter mile to a half mile before the collision. The jurors were especially interested in how the two vessels responded to the rules of passage. The rules, controversial prior to the disaster, favored the *Augusta,* because the *Lady Elgin* was more maneuverable. The weather conditions factored into the judgment of how the rules were applied, as did the extremely perilous sailing condition of the *Augusta.*

The jury's decision was probably predictable. Both vessels were struggling to avoid troughs in twelve- to twenty-foot waves. Nature, far more than human decision, commanded how the lake boats' captains dealt with split-second decisions. Both vessels had been overloaded—the *Augusta* with cargo, the *Lady Elgin* with passengers—which reduced their ability to maneuver under duress. The *Lady Elgin* had a woefully small number of lifeboats to carry passengers in the event of a sinking.

The jury absolved both captains of blame for the collision. The verdict, delivered the day after the accident, found the *Augusta*'s second mate incompetent, deciding he should have been more vigilant after initially spotting the *Lady Elgin*'s lights and not maintaining sound vision of other ships' movements in the storm; in addition, he should have communicated better with Malott. Malott was criticized for not anchoring and assisting the *Lady Elgin* survivors.

The *Lady Elgin* was criticized as well. The jury believed that she should have had watertight bulkheads, that the boat might have sustained buoyancy if only one compartment had been flooded. The boat had been badly overloaded, the verdict lamented, which led to a much higher body count than there might have been.

The main area of blame, the verdict concluded, was neither the boats nor the crews: the biggest problem could be the existing laws of navigation. The use of lights was improper, and the rules of passage needed to be altered. Whether any of this would have mattered in the collision was either disregarded or misunderstood.

This might have been a hearing marred by timing. Emotions ran high when victims' remains were still being collected. All that really mattered, certainly more than assigning blame, had yet to settle on—or in—the lake.

〰

MALOTT NEVER FULLY RECOVERED from the *Augusta*'s collision with the *Lady Elgin* and all the attendant publicity. He continued to sail, but not without difficulties. His reputation appeared to have sunk with the *Lady Elgin*. Sailors were reluctant to join his crews. The superstitious ones, it was said, believed he and his vessel were cursed. The *Augusta*, repaired, repainted black, and renamed the *Colonel Cook*, sailed until 1898, when she grounded on Lake Erie and broke apart.

By then, Malott was long gone. He skippered several vessels, including salties, all for Bissell & Davidson, a small but successful shipping company. While making a delivery in England, he landed a new job when Bissell & Davidson launched a new flagship, the *Mojave*, a bark nine feet longer than the *Augusta* but better designed for hauling cargo. He was given the boat's command when he returned from England.

On September 8, 1864, exactly four years after his collision with the *Lady Elgin*, Malott and his ten-man crew were sailing across Lake Michigan in almost perfect weather conditions. They encountered another vessel, a schooner called the *J. S. Miner*. They acknowledged each other and carried on.

The *Miner* would be the last boat to see the *Mojave* on water. No one could speculate what happened, but the *Mojave*, Malott, and his crew vanished, midlake, without a trace. The wreckage was never located. Malott was last seen alive on the anniversary of his collision with the *Lady Elgin*, on his thirty-first birthday.

〰

THE *LADY ELGIN* STORY did not end with the loss of Captain Malott or, years later, the *Colonel Cook*. The Civil War eliminated the attention on the *Lady Elgin*, but the boat's story stretched on, in various ways, for more than a century. She was remembered in sympathetic press stories published on significant anniversaries. A cathedral in Milwaukee's Third Ward honored the memory of the Union Guard annually on September 8. Sport divers, armed with new equipment that allowed them to stay in the water longer and deeper than ever before, searched for remains of the *Lady Elgin* other than the wreckage washed up on the beach immediately following the sinking; the boat had simply disappeared. Songs were written to commemorate the story.

On May 23, 1989, the 129-year-old mystery regarding the *Lady Elgin*'s whereabouts was solved when Harry Zych, a shipwreck hunter using a side-scan sonar in the area where the *Lady Elgin* was believed to have disappeared, detected something large in about fifty-five feet of water. A dive to the site turned up two large boilers and a substantial debris field. Zych and his fellow divers, hoping they might have located the long-lost wreckage of the *Lady Elgin,* scoured the area for definitive proof. They found it. Searching through the rubble, Zych discovered, among many items, a musket, a sword, a whistle, and a spoon with *Lady Elgin* engraved on it.

This was only the beginning. In subsequent dives moving away from the original debris field, other debris fields were located—fields that helped Zych piece together a sequence of the *Lady Elgin*'s final struggle. None contained large sections of the boat's hull, other than finding a section of the bow, the last piece of the *Lady Elgin* to sink. For all anyone could tell, the vessel had disintegrated over time.

Zych did find many treasures and artifacts. The boat's safe, rumored to have contained a fortune in gold, yielded many gold coins, though not the projected amount. Eleven muskets, probably used by the Union Guard in a Chicago parade and almost certainly purchased by the group to replace those seized by the Wisconsin governor, were found. Four large iron stoves (cargo intended for sale in Superior, Wisconsin), dishes, tools . . . all were found scattered in the debris fields. The *Lady Elgin*'s large paddlewheels offered silent testimony to the breakup of the boat.

Zych brought up approximately two hundred artifacts during the summer of 1989—which did not sit well with the State of Illinois, precipitating a protracted series of legal skirmishes lasting for a decade. The Abandoned Ship Act, passed in 1987, strictly prohibited the removal of artifacts from the remains of abandoned wrecks. Violation was a felony. Zych disagreed with the idea that the *Lady Elgin* was abandoned. Shortly after finding the wreckage, Zych struck a deal with the company that had insured the *Lady Elgin,* an agreement that awarded Zych the boat in exchange for 20 percent of profits from monies earned from the artifacts.

The court proceedings brought into focus the perpetual disagreement over shipwrecks. Divers had plundered wreckage for decades, taking items that might have been appropriate for museums. For their part, ship owners and insurance companies wanted little to do with the high costs of salvaging a vessel that could rust hundreds of feet beneath a lake's surface. Underwater archaeologists, sport divers, salvagers, shipwreck hunters, and anyone else wanting to go to the trouble of locating a wreck and retrieving the artifacts, including pieces of the vessel, were free to do so.

The *Lady Elgin* presented an arguable case. Called "the *Titanic* of the Great Lakes," the *Lady Elgin* wasn't just another unknown shipwreck; she was a part of history, a relic of another era, an icon. The sheer number of people lost made her almost sacred. While perhaps not as infamous as the *Edmund Fitzgerald,* which, in 1975, is the most recent major loss on the lakes and the subject of a song by Gordon Lightfoot, both were subjects of great dispute.

The two battles differed. The *Fitzgerald,* an American-owned boat, had sunk in a violent storm at the cost of twenty-nine lives. No one from the crew was ever recovered. The vessel, to the disappointment of the victims' families, had been explored in her deep resting place. No crewmen were seen during the visits to the wreck, but the crew members' families strongly felt that the freighter was a grave site that should not be disturbed. The *Fitz* had sunk in Canadian water, and Canadian authorities agreed. The wreck, it declared, was off limits.

The *Lady Elgin* presented an entirely different scenario. The boat had sunk more than a century earlier, and survivors and victims' immediate family members had passed away. There were no remains present in the debris field. There was a monetary issue: should individuals be allowed to profit from the wreckage? And where did the State of Illinois, of which a good portion of Lake Michigan was a part, fit in?

Zych and the State of Illinois volleyed in the courts for years. Zych was awarded ownership in one decision, but in the appeal he and the state shared ownership. Other groups, including the Underwater Archaeological Society of Chicago (UASC) and even a representative of the Smithsonian Institution, entered the dispute. Acting on behalf of the state, they explored the debris field and catalogued their findings. It was long, grueling work. The scattered debris, "spread for more than a square mile of lake bottom," made the *Lady Elgin* "the most scattered shipwreck in the Great Lakes."

The years and hearings dragged on. Zych and Illinois accumulated victories and defeats, with still no final, binding judgments. Parts of the dispute reached the Illinois Supreme Court. In 1999, a definitive ruling was rendered, ending a cantankerous decade: Harry Zych was the sole owner of the *Lady Elgin*; anyone or any organization wishing to explore the wreckage was obligated to clear it with Zych prior to doing so.

Throughout the ordeal, Zych had been portrayed both positively and negatively, but Valerie Olson van Heest, who wrote two books on the *Lady Elgin*, dove the debris fields, and was a member of the UASC, probably gave the most even-handed account of the process and Zych's involvement with the *Lady Elgin*:

> In hindsight, it seems that Zych did what he thought was best to protect the artifacts, follow proper legal procedure, and recover a financial reward for his efforts. Unfortunately, the cost of both time and money to mount a ten-year legal battle may not have been worth Zych's ultimate reward.

Alpena (1880)

SEVERE STORMS are legendary on the Great Lakes. They've brought down countless boats since the beginning of commercial shipping. Some have been so memorable that the storms were named after the vessels they destroyed. So it was in the story of the *Alpena,* a passenger boat destroyed by a vicious storm in 1880 that became known as the *Alpena* Storm.

Popular among her regular passengers, the *Alpena* was a Lake Michigan favorite, a beautiful boat that, a saying went, looked "like a white swan on the water." Constructed in 1866 in Michigan City, Michigan, for Gardner, Ward & Gallagher, the *Alpena* featured an attractive double-paddlewheel style preferred by passenger services, with cabins built on a promenade deck and the main deck reserved for shipping freight. At 175 feet, with a 35-foot beam, she was bulky enough to easily accommodate a heavy workload of passengers and cargo—duties performed without event in her early years.

Like most lake vessels of the time, the *Alpena*'s heaviest equipment was carried near the center of the boat. Twin boilers in the engine room belowdecks fed steam to an engine, much of which was mounted outside (known as a "walking beam") and capable of pivoting to supply steam to the boat's paddle wheels.

The *Alpena* was a profitable venture. She sailed along Michigan's western coast until her mission changed to a lake crossing and journey down the western side of Lake Michigan to Chicago. She was sold for $80,000 on April 10, 1868, to the Goodrich Transportation Company. Albert Edgar Goodrich was an ideal owner for the *Alpena.* Born in Buffalo in 1825, Goodrich spent his youthful years in the port city before moving with his family to Michigan. He eventually took a job on the Great Lakes, beginning as a clerk and working his way up to captain. He proved to be every bit as ambitious and skilled a businessman as he had been a sailor. He preferred the side-paddler passenger boats to the straight-deck freighters and schooners of his day. He chartered his first boat, the *Huron,* at age twenty-one, and that same year, 1856, he purchased land near the Chicago River. Over the ensuing years, he expanded greatly, starting the Goodrich Steamboat Lines, buying or leasing a fleet of vessels to operate, and purchasing more property, especially in Wisconsin. His company became established as one of the finest passenger line operators on the Great Lakes.

The *Alpena*

The well-publicized loss of the *Lady Elgin* and the Civil War damaged the passenger business for a few years, but Goodrich's company muddled through without taking too big a financial hit. The loss of the boat the *Seabird* on April 9, 1868, nailed Goodrich Steamboat Lines with its first serious loss. The *Seabird* was relatively new, built in 1859 by R. C. Conwell, and she offered the beauty and amenities passengers had come to expect from Goodrich. On her final voyage, the liner's first trip of the season, the *Seabird* was taking passengers from Manitowoc, Wisconsin, to Chicago. A fatal and easily avoidable error occurred when a porter, taking hot ashes and embers outside, tossed his load overboard. The wind blew it back to the ship. A tremendous fire ensued and could not be extinguished. The captain tried to direct the boat toward shore; the wind fanned the flames. A boat near Waukegan set out to rescue survivors but only managed to find three. About one hundred passengers perished (no manifest existed), marking the *Seabird* as one of the greatest losses yet experienced on the lake.

This unfortunate dramatic and tragic loss of life also resulted in the loss of a vessel, which created an expensive vacuum for Goodrich's company. It was the beginning of a new shipping season, and the fleet was missing an important piece. The *Alpena* became her replacement.

~

WHEN CAPTAIN NELSON W. NAPIER loaded cargo on the *Alpena* on Friday, October 15, 1880, the day was clear and unusually warm. He would be making his customary run across Lake Michigan and down the eastern coasts of southern Wisconsin and northern Illinois, stopping in several ports until he reached Chicago. In uncertain autumn days, this was the type of weather that sailors dreamed of. Aside from an assortment of businessmen, immigrants, and weekenders looking for a late-season retreat, the *Alpena* would be dropping off cargo that included wood stuffing for mattresses and thirteen thousand bushels of apples.

Napier had cause for some concern, however. Barometers indicated a drop in pressure, a surefire indication of an approaching storm. During a time predating any significant national weather service (before radar, jet stream measurements, and storm tracking systems), air pressure, wind speed, and precipitation were about the only measurables at mariners' disposal. Guglielmo Marconi was still more than a decade away from coming up with wireless communications. Sailors might talk to new arrivals on the docks, or captains might get sketchy reports in dock offices, but more often than not they relied on experience when deciding whether to take a vessel out or to stay in. Skippers had to report to offices expecting profits; crewmen had stories to embellish. Virtually everyone had been through an especially nasty storm. Their boats were strong, and their captains were experienced enough to handle whatever Nature threw at them.

But the best prognosticators on Earth could never have predicted the strength of the storm that was blowing in on this day—a national phenomenon that watchers at the time would call "the worst in Lake Michigan recorded history." The storm gained in momentum and intensity as it tracked through the Plains states, stalling over Iowa and Minnesota, dumping snow accompanied by very high winds throughout the area. Temperatures, warm in front of the storm system, fell precipitously. Captain Napier was unaware, as he prepared to leave Grand Haven, Michigan, that the storm would torment Lake Michigan more than any of the other Great Lakes.

The *Alpena* pulled away from the Grand Haven docks around 7:30 p.m. The bad weather had yet to manifest itself, and, in all likelihood, Captain Napier reasoned that he'd have a good jump on crossing the lake before he confronted rough sailing conditions. The storm was intensifying when, at roughly 1:00 a.m., the *Alpena* encountered the steamship *Muskegon,* whose captain would report that all seemed well. The two vessels saluted one another and sailed on into the dark, overcast night.

The wind picked up within an hour or two, roaring in from the southwest. The *Alpena* was now facing gale-force winds and very high

The *Alpena* amid the storm

seas. At some point—and no one would ever know for certain when or where—the *Alpena*'s fight with the elements reached a tipping point. She was on her own, facing what would turn out to be a losing battle. She wouldn't reach Chicago.

~

GOODRICH'S CHICAGO OFFICES displayed no concern when the *Alpena* failed to turn up on schedule. Vessels commonly went missing during violent storms; they would seek safe harbor in some port or, quite often, hide behind an island or somewhere else in the lee of the storm. They didn't possess the equipment necessary to contact their offices about their whereabouts. They would simply disappear, resume sailing when the weather permitted, and report to their designated harbor as soon as possible.

Goodrich officials took for granted that this was the explanation for the *Alpena*'s tardy arrival. When all the numbers were added up, the storm had taken an extremely harsh toll: ninety-four vessels had been lost or badly damaged. The storm still raged on Saturday. The wind screamed, leaving a trail that broke windows throughout

Chicago. Temperatures dropped to well below freezing, and snow fell. The Grand Haven–Chicago route took a day and a half under good sailing conditions, so there was no reason to worry when the *Alpena* didn't report. The boat's sixty-four-year-old captain, who'd earned his master's license forty-six years earlier, knew all the procedures. Besides, three other Goodrich Line vessels, the *Menominee, De Pere,* and *Corona,* were also late and unreported.

The concern began on Sunday. The storm headed east, and boats checked into the Chicago Harbor. Reports from crews from other vessels—unsubstantiated but not to be ignored, either—were not encouraging. The nearest to Chicago the *Alpena* had supposedly been seen was about thirty-five miles off Kenosha, a city on the Wisconsin–Illinois border. The boat, said the captain who had seen her there, was fighting to stay afloat. Another captain said he'd seen the *Alpena* on her side, with one paddle completely out of the water, adrift in the storm.

Captain John Dearkoff, piloting the *Hattie Wells,* offered the most harrowing, detailed eyewitness account of a boat he thought was the *Alpena.* The *Wells* was taking a terrible beating and facing the possibility of capsizing when Dearkoff ordered his wheelsman to make a run for Milwaukee. They would wait out the storm in the safety of the city's harbor. "I don't know why he didn't head to Milwaukee like I headed for Milwaukee," Dearkoff said of the man he thought to be the *Alpena*'s Captain Napier. "Me and the crew stood on deck and watched him try to turn his ship around in that storm. She was halfway around and that wind just took right hold of her and turned her over. She was swamped and she sank and we couldn't do a thing about it. We watched her disappear under the waves."

The fate of the *Alpena* was confirmed when wreckage washed ashore over the following week—and beyond. Fire buckets with the boat's name stenciled on them wound up on the beach, as did a large portion of the *Alpena*'s piano, deck chairs, a cabin door, a piece of the vessel's grand stairway, life preservers, shoes, coffins, and small pieces of torn-away wooden wreckage. Only a few bodies found their way to shore, and two wore watches that had stopped at 10:55. This offered nothing conclusive: the time was now known, but not the date.

Though answering the question of whether the ship had survived, the bodies and wreckage presented many more. Those who reported seeing the *Alpena* had done so near Wisconsin. The victims and wreckage had all washed ashore on the Michigan coast, particularly near Holland. Over the ensuing weeks, apples bobbed on the lake's surface in the same area. Observers agreed that this seemed to indicate that the *Alpena* had sunk near Michigan, not far from where she had originally set off. The recovery of the vessel's piano underscored this speculation. "It being barely able to float, our sailors concluded that she did not come very far," said one observer. "And the arrival of other heavy pieces of the wreck would seem to corroborate this."

But what about the *Alpena*'s sister ship spotting and signaling her in the middle of the lake? The storm's winds had shifted, and heavy winds and seas had pushed vessels across the lake in the past, but this was hard to accept in the case of the *Alpena,* given all the conflicting testimony. Could it be that Dearkoff did indeed see the *Alpena* turn around, but rather than sink there, as he insisted, she had sailed back toward Michigan, only to sink before reaching the state's coastline?

The *Alpena* had sunk—no one doubted that. But where, and when? And how? Was she in shallower water near shore, or had she come to rest farther out, in deeper water? And how many people had been lost?

One piece of cryptic evidence about the boat's final moments found its way to land in the form of a water-soaked note nailed to a piece of cabin molding:

> Oh! This is terrible. The steamer is breaking up fast. I am aboard from Grand Haven to Chicago.

The note was signed by either George Conner or George Connell. Water had made the last letters of the signature illegible, and since no one came forward with an inquiry about a passenger with a similar name, the note, too, became a mystery.

The already muddy waters of speculation were clouded further when, a short time after the first discovery of wreckage, the *Alpena*'s 5-by-10-foot flag drifted to land near White Lake Channel. This Michigan location was quite a distance from where the other wreckage had been found.

No shortage of theories purported to answer these and many other questions. An assortment of ideas addressed the question of *how* the *Alpena* had sunk. The most widely accepted explanation claimed that the apples, piled on deck, shifted in the roiling waters and upset the *Alpena*'s stability, rolling her onto her side and ultimately sinking her. Variations of the shifting cargo theory abounded. Another theory was more basic: caught in a trough and unable to escape, the *Alpena* was overwhelmed by huge waves and foundered. This appeared to endorse Dearkoff's account. Still another theory supported the brief note found on the Michigan beach: the storm had beaten the *Alpena* into pieces so that she could not survive. There was speculation that the boat had collided with another vessel.

The idea that the *Alpena* may not have been seaworthy at the time of her sinking was also bandied about, creating a controversy that Goodrich Lines vigorously denied. The *Alpena* had undergone minor repairs before her final voyage, and Goodrich was known for the maintenance of its fleet. An article in the *New York Times,* on October 22, 1880, reported that Michigan observers onshore had deemed some of the timbers drifting to the beach "absolutely rotten" and declared that they "considered [it] remarkable that she had not gone to pieces earlier." The *Cleveland Herald* published a piece in early 1881 mentioning that a coroner's jury had judged the *Alpena* to be unseaworthy, "with inadequate and worthless life preservers, with inexperienced sailors for a crew, and with at least one rotten lifeboat." The writer did not detail how the jury reached its conclusion.

A. E. Goodrich, incensed by the charges and their implications that Goodrich Lines was negligent in any way, fought back. The rotten, waterlogged wood, he argued, might not have been from the *Alpena*; the violence of the storm had stirred up the water to such an extent that the wood could have originated from many sources. (One lighthouse keeper, in a statement unrelated to the coroner's jury, said that the water near Door County, Wisconsin, had been so badly stirred up that the water in the area remained white for a week after the storm; this, he claimed, was due to the limestone on the lake bottom's being stirred up and mixed with lake water.) Goodrich countered that the jury's charges were "unsupported by the facts, and the conclusion that company shall be held responsible for all damages arising from the disaster [were] unwarranted."

The bickering continued for months, with no definitive conclusion drawn. The exact location of the *Alpena* remained unknown. The vessel was gone, relegated to ghost ship status. Between 50 and 110 passengers and crewmen (depending on whose estimates you believe) lost their lives. The Goodrich company concluded, despite evidence to the contrary, that the *Alpena,* caught in a trough, had sunk closer to Wisconsin's shoreline than to Michigan's. The storm, eventually called "the Big Blow of 1880" and "the *Alpena* Storm," covered more than five hundred square miles of the Midwest and Upper Midwest and marked the beginning of one of the worst winters the area had seen up until then.

Twenty-nine years after the vessel's loss, on the heels of a January 1909 storm, a large piece of wood was found on shore near Holland, Michigan. The wood had the name *Alpena* written on it. This turned out to be the nameboard from one of the ship's paddlewheels.

Appomattox (1905)

INNOVATION WAS THE WORD when, in 1896, construction of the largest steam-driven, wooden-hulled freighter in Great Lakes history commenced in West Bay City, Michigan. Shipbuilders no longer favored wood—the lucrative iron ore transport business, thriving since the discovery of gigantic deposits in Minnesota, had seen to that. Wooden schooners and freighters were less capable of withstanding the beatings those loads administered than were the sturdier freighters constructed of steel. There was also the issue of better fireproofing for the vessels. Boilers, the root of so many conflagrations on the lakes, were improving in design, but they were far from fail-safe. Wood, in the age of steam-driven freighters, seemed to invite trouble.

The freighter under construction in 1896, to be christened the *Appomattox,* was the brainchild of renowned shipbuilder James Davidson, who not only designed and built wooden-hulled boats but owned his own shipping firm. The Buffalo native launched Davidson Shipbuilding in 1873 and one year later launched his first boat, the *James Davidson,* which was followed by many others. He preferred to work in wood and was said to be the first to fashion wooden hulls cut specifically in sawmills. Later in his life he would become a highly successful banker.

A vessel as large as the 319-foot *Appomattox* required inventive fortification of its oak hull. Huge steel arches and cross braces added strength to the hull, as did the inclusion of steel keelson plates. Davidson's master plan involved the *Appomattox* hauling a fully loaded schooner barge behind her. The two vessels could handle larger cargos efficiently; they would haul iron ore from Minnesota to Michigan and then return with full loads of coal.

The *Appomattox* was a family venture. Davidson owned the freighter and the company bearing his name. His son-in-law, G. A. Tomlinson, vice president of the company, worked as the boat's master in the early years of her operation. The company eventually relocated to Duluth, which then became the boat's home port.

The *Appomattox* functioned as designed without serious incident, with one exception. On August 3, 1900, while sailing on the St. Clair River with her usual consort, the wooden schooner *Santiago,* the *Appomattox* hit a thick patch of fog. Another vessel, the *Kaliyuga,* also towing a schooner, the *Fontana,* was approaching. The *Santiago* veered off course and collided with the *Fontana,* tearing a gaping hole in her hull. The schooner sank quickly, and one man died.

〰

IT LOOKED LIKE another ordinary, uneventful run on November 2, 1905, when the *Appomattox,* loaded with 3,443 tons of soft coal, left to head down Lake Michigan with the *Santiago,* carrying 4,735 tons of coal, in tow. As any mariner knew, November was the most challenging month of the year in terms of sailing conditions, and this one would be especially so. For the time being, though, the lake was in fine sailing form.

This changed when the two vessels reached southeastern Wisconsin, not far from Milwaukee. A dense fog combined with industrial smoke. Trying to find her way in the thick air, the *Appomattox* ventured too close to shore, and she and the *Santiago* grounded in

The *Appomattox*

the shallows, the contact with the rocky lake floor damaging the bottom of the *Appomattox.* Another vessel, the *Iowa,* sailing from Sturgeon Bay to Milwaukee, suffered the same fate nearby. Captain John Rawley of the *Iowa* blamed the thick air for his accident. "There was a heavy fog offshore this morning," he told a reporter from the *Milwaukee Journal.* "It was this that caused us to run ashore."

The crews from all three vessels were helped safely ashore. No one was injured. Two tugboats, a revenue cutter, and men from the local U.S. Life-Saving Service station quickly went to work trying to free the three boats. The *Santiago* and *Iowa* were refloated without much difficulty. The same could not be said about the much larger *Appomattox.* The boat was breaking apart and was laden with a full load of coal. Crews were able to dump most of the cargo, but they had trouble pumping out the water that had entered the ship after the grounding. The weather worsened, kicking up high seas. The crews worked through it until, on November 15, they surrendered to the obvious. The *Appomattox* was declared a total loss. Talk turned to salvaging the boat's engine and boilers.

All attempts to salvage the *Appomattox* ceased after one of the most powerful storms to sweep the upper Great Lakes bore in on November 27, causing extensive damage to the area. Known today as the *Mataafa* Storm, after a giant freighter that became stuck outside the Duluth harbor, costing all the men trapped in the aft section of the vessel their lives, the storm blew onto Lake Michigan and worked its way southward. Most captains wisely chose to keep their vessels in the harbors around Lake Michigan, avoiding the human and material loss that plagued Lake Superior.

Between the frigid temperatures and rough water, the winter following the storm broke the *Appomattox* in three pieces and sank her in fifteen to twenty feet of water. Her coal cargo washed ashore for years, along with bits of wreckage. In 1907, the Reid Wrecking Company of Sarnia, Ontario, successfully removed and brought up the engine and boilers. The rest of the wreckage settled and compacted over time. Its proximity to shore and its shallow depth made it a popular site for amateur divers, who could see sections of the hull, along with a nearby propeller; underwater archaeologists explored it for clues about the transition between wooden and iron crafts. Every summer, a buoy marking the boat's location was set out on the lake.

The *Appomattox* wreckage was listed on the National Register of Historic Places in January 2005. The vessel's life was relatively short, and her end was unspectacular, but she remains noteworthy in Great Lakes history.

The propeller of the *Appomattox*

R. J. Hackett
(1905)

COMMERCIAL SHIPPING evolved exponentially during the 1800s, from the types of vessels to the materials used in building them to the cargos headed from port to port. The very first commercial endeavors had involved birchbark canoes that traveled little distance from the safety of shore; the nineteenth century opened with wind-driven wooden vessels capable of crossing a large body of water and, if necessary, confronting the challenging conditions of Nature. Land was being quickly developed, and the population was growing. Everyone needed day-to-day provisions to keep them fed, clothed, and protected from the elements, but the expanding cities required building provisions as well.

The number of commercial vessels increased. They grew in size, with larger bellies to hold larger cargos. Wood was plentiful; there was no shortage of boat construction. Elongated two- and three-masted schooners, their billowing sails powering them through water at unprecedented speeds, dominated the business, joining and then surpassing vessels modeled after their European ancestors. But the reliance on wind and seamanship presented scheduling issues.

The invention of the steam-driven engine altered that. Boats with engines and boilers, while requiring new skills from crewmen, began replacing vessels dependent on wind. Predictably, new problems arose. Imperfect boilers exploded, causing fires on flammable carriers. Scores of boats caught fire and burned to the waterline. Hundreds of people lost their lives. Safety issues forced the hands of politicians and shipping company officials.

Then came the introduction of vessels constructed of iron and steel. These started appearing in the latter part of the nineteenth century—and not a moment too soon. Enormous deposits of iron ore had been discovered in Minnesota and, to a lesser extent, Wisconsin, and with the growing demand for iron came the need for larger, stronger boats to haul it.

The *R. J. Hackett* seemed to embody this overall evolution. Launched in 1869, the 208.1-foot boat, though constructed from wood, offered many firsts (or near firsts) that helped shape the future of commercial shipping on the Great Lakes. She was the first vessel built with the iron ore (and grain) business in mind. Her design, unique at the time, would serve as a template for the design of boats for the next century; she was one of the first boats to tow a consort behind her, opening up possibilities previously unheard of in the amounts of cargo moving between ports.

She might not have been the biggest or best, and even her demise in 1905 was overshadowed by the great storm that destroyed so many vessels on Lake Superior. But there was no overlooking her overall importance. For thirty-six years, she was a standard-bearer.

~

THE *R. J. HACKETT* was the brainchild of Elihu M. Peck and the product of creativity meeting necessity. Peck packed a lot into his seventy-three years. Born on September 14, 1822, in Butternuts, New York, he moved to Cleveland in 1838 and began his maritime career as a ship's carpenter. He also worked as a shipbuilder's apprentice and, while still in his twenties, formed a shipyard and began repairing and

The *R. J. Hackett*

building his own vessels. He consumed anything he could learn about Great Lakes shipping, including how to run a boat. He obtained his master's license for a brief time, piloting the *Fountain City,* which he owned.

Peck met and grew close to shipbuilder Irvine Masters. The two worked together and, in 1854, founded the firm Peck & Masters. Shipping on the Great Lakes was enjoying an upswing, and the firm grew with it, using its founders' ambition and innovation to become one of the most successful enterprises in the business. Masters suffered from poor health, though, and died in 1866, three years prior to the construction of the *R. J. Hackett,* but Peck kept the firm operating.

The construction of the *R. J. Hackett* combined Peck's various maritime experiences and ideas. With the new and growing demand for as much iron ore as could be shipped, foremost on Peck's mind was a radical change in design. By tradition, bulk carriers were constructed with heavy machinery in the middle of the vessel and the pilothouse above deck—almost exactly above. This was efficient when the boat was delivering such freight as lumber or grain, but in Peck's mind it was wasting valuable space in the iron business, in which packed cargo holds meant greater earnings. Profits depended on tonnage.

For the *R. J. Hackett,* Peck relocated the pilothouse and officers' quarters to the bow of the vessel, which gave an added advantage of better visibility; the engine room, galley, and other crew quarters were moved to the stern of the boat.

This design left the middle of the boat with more room than ever for cargo—which, after all, was the reason for the carrier's existence. The *R. J. Hackett*'s cargo hold, boxy in appearance, was designed to carry twelve hundred tons of iron ore, loaded through evenly spaced hatches cut into the deck. Peck's hatch design matched the loading dock at Marquette, Michigan, one of the leading ore docks on the lake. But this was not all. To haul extra tonnage per trip, the *R. J. Hackett* was set up to tow a consort behind her. Another vessel, the *Forest City,* was created for this purpose. The 216-foot schooner barge was, in some ways, similar to the *R. J. Hackett,* but she was constructed without an engine.* The *R. J. Hackett* began towing her in 1870.

These design innovations, considered radical at first, elicited mixed responses from the shipping industry. The boat looked strange, especially with her tall twin masts, positioned at either side of the main deck and used with sails to add power and stability to the vessel. The lengthy deck and cargo hold required longitudinal reinforcement. When Peck launched the *R. J. Hackett* and put her up for sale, he could find no buyers. He and his investing partner, Robert J. Hackett, founded the Northwest Transportation Company and kept the boat for themselves.

~

THE FIRST SERIOUS PROBLEM of the *R. J. Hackett* on the Great Lakes occurred in 1871, the same year she set a mark as the first boat to deliver iron ore to Cleveland. On May 27 of that year, she grounded in the mud and silt near the St. Clair Flats, where she remained stuck for nine days. The area was a trap for boats entering it, but it was especially treacherous when water levels were low. "Lake levels were

* An engine was added to the *Forest City* in 1872.

lower than they had been in years," reported newspaper columnist James Donahue, noting that "grounding was a common problem that year."

A bad situation was magnified when another boat, the *Dean Richmond,* tried to find a way around the disabled *R. J. Hackett.* It, too, grounded, completing a blockage of the waterway. All traffic halted. Ships awaiting passage backed up. Attempts were made to free the two grounded vessels. Cargo was removed from the *Dean Richmond,* the idea being that a lighter vessel might be pulled from the mud. The efforts paid off, and the *Dean Richmond* was freed after a day's work.

The *R. J. Hackett* was a different matter. Sailing without cargo, she and her consort, the *Homer,* could not be immediately lightened. For days, men labored to free the *R. J. Hackett,* first by removing any weight and then by employing hawsers to try to attempt to pull the ship free. The boat remained stuck and the stalled lake traffic continued to pile up. Finally, the Coast Wrecking Company, a salvage firm, dug in the mud around and in front of the vessel and attached pontoons to her. On June 5, the *R. J. Hackett* was buoyant again.

Any relief that Captain Trotter and his crew might have felt ended only a few days later when the *R. J. Hackett,* engulfed in dense fog while sailing to Escanaba, Michigan, grounded again, this time near Thunder Bay Island. It took four days for tugs to pull her free.

The sailing history of the *R. J. Hackett* was otherwise unremarkable but very profitable. The boat soldiered on while other shipbuilding companies copied her design, though steel was replacing wood as the material of choice in boat construction. In 1881, the *R. J. Hackett* was overhauled and lengthened to 211.2 feet. She received a second deck, and any aging timber was replaced, all at a cost of about $8,000. Two years later, she received a steeple compound engine.

By the turn of the twentieth century, the best days of the *R. J. Hackett* were behind her. She was still seaworthy and in good repair, but she was also one of the last of the wooden bulk carriers operating on the lake. Ownership changed hands in 1892 and again in 1905, winding up with C. C. McCallum, who invested nearly all of his assets in the vessel. McCallum, a captain, guided the *R. J. Hackett* for the season. By all indications, it was a profitable one. It would be McCallum's only season as the vessel's owner and skipper.

THE 1905 SHIPPING SEASON was drawing to a close when McCallum and his crew of fifteen departed for the Marinette Fuel and Dock Company with twelve hundred tons of coal. McCallum was seasoned enough to know all the temperamental November sailing conditions on the Great Lakes, but apart from the seas running high, McCallum saw nothing to indicate a trip out of the ordinary. Indeed, the weather would not be an issue on this, the final voyage of the *R. J. Hackett.* The boat was sailing along smoothly, on schedule, as she neared Green Bay, Wisconsin.

It was breakfast time in the galley on November 12 when a fire broke out in the crew's quarters. It spread quickly. The crew initially believed they had extinguished the fire, but the men were mistaken. Remnants of the fire were found near the engine room, and the stern section of the *R. J. Hackett* was soon engulfed in flames. McCallum ordered his wheelsman to ground the boat on Green Bay's Whaleback Shoal.

The crew, seeing the *R. J. Hackett* in mortal danger, abandoned their firefighting and launched two lifeboats, with everyone, including McCallum, aboard. "Just as we left the boat the stack toppled over and struck the whistle rope and the whistle blowed steady until the craft blew up," Captain McCallum told the *Marinette Eagle.* "The coal in the hold of the boat produced so much gas that the whole upper works went up with a report like a gun."

The *R. J. Hackett* was still burning when two vessels came to her assistance. The small crew of the fishing tug *Stewart Edwards,* out checking nets, caught sight of the *R. J. Hackett* and lit out after her. Then Island Lifesaving Service's lookout and the lighthouse keeper saw the conflagration and the thick black smoke rising from it. The lifesaving crew responded but found the high seas and considerable distance from the *R. J. Hackett* a formidable challenge. By the time they reached Whaleback Shoal, the fishing tug had already rescued the crew of the *R. J. Hackett*, and the vessel was a smoldering mess. The crew managed to board the boat and salvage some of Captain McCallum's papers and personal effects.

The *R. J. Hackett*, officials decreed, was a total loss.

THE LOSS OF THE *R. J. HACKETT* made news throughout the region. She had been the first of what became known as Great Lakes freighters, and though she was no longer the biggest or strongest of these boats, she was historically significant, the prototype of almost all of the boats constructed during the following century. The story of the *R. J. Hackett* did not dominate shipping news for long, however. One of the most powerful storms in the annals of the Great Lakes—one that sank and destroyed dozens of vessels and claimed many sailors' lives—hit Lakes Superior and Michigan only a couple of weeks later. The *R. J. Hackett* became a memory, a notable footnote.

The *R. J. Hackett* wreckage slipped from the shoal and settled in shallow water, the top of it in just over ten feet of water. Divers visiting the site in the late 1980s found the wreck in better-than-expected

A map of the wreck site of the *R. J. Hackett*

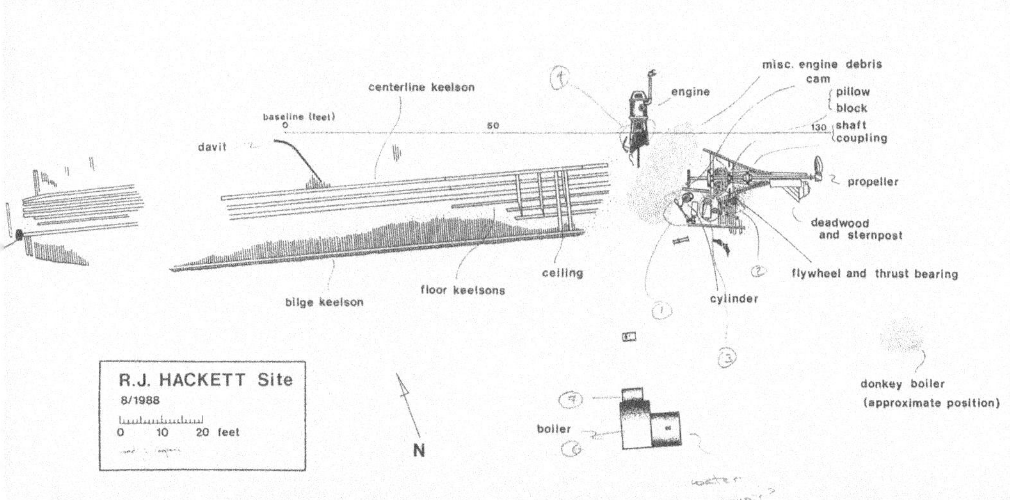

condition, with much of the unburned portion of the hull relatively intact. Scavengers had removed some artifacts from the wreck, but others remained, such as the propeller shaft, portions of the engine, a boiler, and smaller items. Writing about the "great historical significance" of the *R. J. Hackett,* the Wisconsin Historical Society recommended further serious exploration of the wreckage, calling it "of considerable appeal to the beginning or intermediate sport diver."

The *R. J. Hackett* was placed on the National Register of Historic Places in 1992.

Salvaged machinery wreckage from the *R. J. Hackett*

Pere Marquette 18 (1910)

JULY 23, 2020

The shipwreck hunters sailed the 33-foot 1947 Owens Cruiser named Heyboy *east from nearby Sheboygan into the open water. Their target: a car ferry, the* Pere Marquette 18, *that had disappeared without a trace 110 years ago in the vicinity.*

It was the kind of ghost ship that attracted shipwreck historians—an "obsession," said Jerry Eliason, one of those aboard the Heyboy *willing to gladly trade whatever frustration they might experience in going out looking and coming back empty from the sheer elation they felt when locating and identifying a long-lost vessel. Eliason and his friend, Ken Merryman, had been at it for years, partnering for numerous discoveries of historical import.*

On August 21, 1977, they'd found the Kamloops, *a 250-foot freighter lost on December 7, 1927, near Isle Royale. The boat had been caught up in a terrible storm, and when she failed to arrive at her destination, the search was on. Despite an intricate scouring around the island, nothing was found—no wreckage, no survivors, no victims. The following spring, searchers found the remains of many of the boat's crew on the island, where they'd found refuge, only to die from the brutal conditions that followed. There seemed to be no reason for the ship not being discovered. but that was how it was for this site, which was "more shrouded in mystery than any other wreck at Isle Royale." Merryman, Eliason, and others researched the boat's story and eventually located the wreck.*

On May 24, 2013, they hit paydirt again when they found the Henry B. Smith, *a brand-new freighter that had gone missing during the fearsome 1913 hurricane on the Great Lakes. This was a prize find in shipwreck hunting history. The boat had left port despite the reservations of her captain, which were ignored by the freighter's owners. Witnesses observed the Smith's departure from harbor and subsequent battle with the elements, and she was apparently running for safety when she disappeared from sight. Only wreckage that washed ashore confirmed what others had feared. No one knew her location until Eliason, Merryman, and others found her nearly a century later.*

The Pere Marquette 18 *had her own unique story. She'd been specifically designed to ferry railroad cars across Lake Michigan, usually from Milwaukee to Ludington and back, but in her last years of operation, she'd spent the summers doubling as a luxury liner—a duty that found the boat stripped of working parts, rebuilt for dancing and partying, and rebuilt again in the fall for more hauling until the following summer. She'd also served as an icebreaker in the winter months. There were those who believed that she had quite literally been worked to death, the victim of too much work and perhaps not enough maintenance.*

The Pere Marquette 18, *loaded with boxcars and passengers, left Ludington for Milwaukee in early September 1910. Somewhere in the middle of Lake Michigan, a crewman spotted water pouring in belowdecks. The crew attempted to stave off the leakage, to no avail. The boat sank suddenly about fifteen miles east of Milwaukee, taking all her officers and a number of crewmen and passengers with her. Her whereabouts remained unknown for more than a century.*

Now, on this nice July day in 2020, shipwreck hunters were going to try to find her. Equipped with a side-scan sonar, an experienced group of searchers, and decent weather, Merryman and company decided to try a different approach. Merryman had gone out to look for the vessel a year earlier, using research based on reports on the sinking, but he'd come away unsuccessful. He knew the drill all too well: contemporaneous accounts from officers, survivors, and rescue personnel were often inaccurate, and in a large body of water the slightest mistake can leave you miles away from your target.

The previous winter, Eliason had pinpointed a new source: the United States Life-Saving Service, active in attempts to find and rescue survivors of the Pere Marquette 18 *sinking, had offered an eyewitness account of its efforts. The account was contained in a federal report, which Eliason was able to obtain. The report had the boat sinking in a location that differed from the research the shipwreck hunters were exploring. Eliason and Merryman discussed various search possibilities and came up with a new plan.*

With weather conditions favorable, they were going to look anew.

GREAT LAKES SHIPPING, from its earliest days, has been the least expensive method of moving bulk freight around the upper Great Lakes and Midwest regions of the country. Still, there were issues that stretched out time and sent schedules off-kilter, and that, of course, translated into lost revenues. A trip from Minnesota to, say, Cleveland involved a long journey down Lake Michigan and time spent in the congestion of Chicago boat traffic. Railways were limited by the same problem.

Competition between carriers on land and sea grew much more intense with each passing year. Railroad traffic increased with the constant expansion of new railways, affecting the maritime movement of passengers and freight. This only served to underscore the bottlenecking of traffic at the bottom of Lake Michigan. Schedules had to be met, people had to be moved. Train service continued to chip away at shipping on water, and railroad officials plotted ways to strengthen the companies' impact.

The solution dredged up an old cliché: If you can't beat them, join them. The brainstorming for officials reached an almost ridiculous solution. Rather than transporting cargo down the eastern portion of Lake Michigan, around northern Illinois, and back up the western portion of the lake on the coast of Michigan, why not ship loaded boxcars across the lake? One could move cargo from Milwaukee, Wisconsin, to Ludington, Michigan, and save time and money in doing so. The idea of the car ferry was hatched.

Car ferries, of course, needed to be large and strong. They had to be powerful enough to move across the lake at a good speed. For additional income, they could be fitted with passenger quarters. They had to be able to sail year-round, which meant they had to be capable of breaking through the ice that accumulated on the lake when winter temperatures plunged. And not to be overlooked or downplayed, it was crucial that the vessels be able to accomplish all this at a reasonable price.

The design of the car ferry was not all that different from the design of turn-of-the-century bulk carriers, with one notable addition: railroad tracks. The tracks would be secured to the decks for the loading and unloading of boxcars at railroad facilities already existing at ports. The cars would be coupled to awaiting trains, with the cargo shipped to scheduled stops at stations on already existing routes.

The first company to put all this together was the Pere Marquette Railroad. PM, as the company was called, was formed when three railroads (the Chicago & West Michigan, the Flint & Pere Marquette, and the Detroit, Grand Rapids & Western Railway) merged in 1874. After chartering boats to haul cargo across Lake Michigan in the long-established way, PM built and launched its first car ferry, the *Pere Marquette,* in 1897. Other car ferries, all christened *Pere Marquette* with a subsequent boat number, followed. The ferries enjoyed success, and PM became established as the premier car ferry builder and operator on the lakes.

Construction of a grand, spanking-new car ferry, designed by Robert Logan and built by the American Ship Building Company of

The *Pere Marquette 18* passes the State Street Bridge in Chicago.

Cleveland, commenced in 1902. At 338 feet, with a 56-foot beam, this latest addition to the fleet—the company's flagship—was a vessel that would require a large crew to handle cargo and passengers. She ranked among the most beautiful and spacious boats on the lakes. With fifty huge staterooms, plus an assortment of smaller quarters, the black-and-white painted vessel could transport hundreds of guests, all of whom were assured of superb accommodations.

Equal attention was given to the working aspects of the craft. Four sets of railroad tracks, capable of holding up to thirty cars, occupied the main deck. For power and speed, the boat contained twin steam-passed, triple-expansion engines, fed by six Scotch boilers, making the vessel one of the fastest on the Great Lakes. She would typically travel at fifteen to sixteen miles per hour.

The *Pere Marquette 18* was launched on August 16, 1902. In his book *Steel on the Bottom,* Great Lakes historian Frederick Stonehouse judges the vessel's christening, performed by the designer's daughter,

Beatrice, to have been ominous. By tradition, a bottle, usually Champagne, was smashed on the boat's hull; in the case of the *Pere Marquette 18*, Beatrice Logan followed a Japanese tradition of releasing doves, which Stonehouse deemed to be insulting to the old salts plying their trades on the Great Lakes.

"There was only one way to christen a ship," Stonehouse wrote, "and it meant breaking a bottle of hooch on the bow and had nothing to do with fowl polluting birds."

"Our perception on where things are changes over time," Ken Merryman said of a searching process that sometimes took years and usually involved eliminating possibilities through trial and error.

Such was the case with the Pere Marquette 18. *Merryman, Jerry Eliason, and others had been searching, off and on, for years—sometimes together, sometimes separately. Based on their research, they would discuss where they might look. There were inevitable disagreements, which would be hammered out in due time. They reached some of their destinations within a rather narrow window of hunting possibilities. Searching was at its most productive in the early summer, when the weather was more conducive to side scanning and photographing. Wind and stormy weather wreaked havoc on both.*

The men felt the weather effects while looking for the Pere Marquette 18. *Eliason and Merryman agreed that the sunken wreck was farther south than they had initially believed. When talking to investigators soon after the sinking, wheelsman Simon Burke, who had been directing the boat west with the hope of beaching her before she sank, spoke of being ordered to turn to the south as crewmen tried to lighten the load by dumping cars overboard. This would have placed the* Pere Marquette 18 *somewhere between Sheboygan and Port Washington, Wisconsin, not due east of Sheboygan as had been believed.*

Eliason and Merryman had differing thoughts on how far out on Lake Michigan the vessel had been when she suddenly sank, stern first, beneath the waves. Eliason felt she had sunk about thirty miles from shore, and he chose to look there on his own, without Merryman. He set a date for the search but was "weathered out" by wind and choppy seas.

This proved to be fortunate. Merryman had based his selected location on several key factors in the federal report. He tried to establish a timeline and use it to figure location. The shipwreck hunters already knew approximately how long the boat had been sailing from her departure in Ludington at the time of her sinking. The new information that the Pere Marquette 18 *had turned southward altered their previous perceptions. The final key factor involved the time it took one of the rescue boats to transport survivors from the site of the sinking back to Ludington. Merryman and Eliason discussed the math, and Merryman marked an* **x** *on a map. This was where they would sail.*

IN THE AFTERMATH of the loss of the *Pere Marquette 18,* questions would arise over whether the boat had been overworked, sacrificing good, regular maintenance in exchange for profits—whether the day-to-day grind had exacted a price on the boat's structure. This was not an uncommon line of questioning when otherwise seaworthy vessels were lost, especially when the loss of life was involved.

The *Pere Marquette 18*'s work patterns changed drastically in March 1907, when the Chicago & South Haven Steamship Company contacted the Pere Marquette Railroad with an intriguing proposal. The *Eastland,* a very popular passage liner offering pleasure cruises out of Chicago, was relocating to Lake Erie, leaving a huge hole in the lucrative cruise business. Was it possible, Chicago & South Haven officials wondered, to charter the *Pere Marquette 18* during the summer months to take the place of the *Eastland*?

There were pros and cons to consider. The profits, of course, were the biggest positive. On the negative side, the *Pere Marquette 18* would have to undergo extensive renovations. She might be fine for her usual load of passengers, but converting to a pleasure cruiser handling thousands would be a costly maneuver. Another deck to accommodate dancing and a large, ritzy palm garden would have to be created. New entrances for those using the dance and orchestra area had to be built. Two new staircases were planned. Finally, as a safety precaution, five thousand life jackets were purchased.

All this was preparatory work (at a cost of ten to fifteen thousand dollars, a very large sum at the time) and it was being done for a limited-time job. When the cruise season ended, the new deck would be removed and stored, along with other equipment used only during the season. Every year, the *Pere Marquette 18* would be converted back into a car ferry and resume heavy duties in hauling freight, railroad cars, and passengers.

Pere Marquette Railroad authorities weighed the pros and cons and judged the cruise business to be worth the investment. The boat's schedule was basic: five days a week, from Monday through Friday, the *Pere Marquette 18* departed Chicago at 9:00 a.m., took a leisurely cruise north to Waukegan, Illinois, and headed for home at 4:00 p.m. Once back in the Windy City, the boat took passengers on a moonlight cruise around the Chicago area. On Saturdays, the vessel stayed close to home, sailing around Chicago. Sundays were reserved for a trip to New Haven. The company spent a princely sum on advertising and promotion, to great effect: the boat was called the "world's greatest excursion steamer."

The *Pere Marquette 18* elicited gushing reviews from the press, as well as from passengers. At the conclusion of the vessel's first season as a cruise ship, the *Ludington Chronicle,* accustomed to observing the boat in another capacity, called her "nothing more nor less than a mammoth floating amusement garden, a summer resort which has conveyed its patrons over miles of dancing waters."

There is no definitive method of measuring how—or whether—the heavy workload affected the boat. Did her function as an icebreaker during the winter season weaken the boat's hull at a time when other vessels received maintenance at dry dock? What effect, if any, did the back-and-forth work on converting the vessel factor into it? Word circulated that the summer crew was hard on the boat, beyond the customary wear: how did this matter in the big picture?

These and other questions would have been difficult to answer, if they were even considered. But the *Pere Marquette 18* was handling her duties without any obvious problem, passing regular inspections and making money.

As the saying goes, why fix what isn't broken?

~

When the Heyboy *arrived at the search location, the search appeared to be in jeopardy. The weather was changing. The wind was increasing, and the seas were growing choppy. All the research, Ken Merryman said, had placed the shipwreck hunters "in a tight pattern of searching," but from the onset, when the side scan was placed in the water and they viewed the underwater topography of Lake Michigan, the men on board wondered whether they might be better served by packing it in and sailing back to Port Washington. The search was tedious enough without being twisted by the elements.*

The actual process was basic, often compared to a groundskeeper mowing grass. A search boat would go back and forth, back and forth, over a given expanse of water, then change direction and repeat the process. The side scan would view the bottom of the lake and send signals to a monitor on the boat on the surface. It was a lengthy, usually unrewarding process until the scan picked up an image or shape of something large. If the scan located something in an area where the searchers believed a sunken vessel was located, the excitement in the wheelhouse was almost palpable. A camera might be lowered—or, if the water was shallow enough to swim down to explore it, the task of identifying the wreckage began.

Depending on the sunken vessel, identification could be easy or difficult. More recent wrecks were easy to identify, as a camera or diver could pick up the name emblazoned on the boat. Older boats—vessels that had been submerged for fifty years or more—could be problematic. Their names might be obscured by the proliferation of quagga or zebra mussels, which could engulf a sunken vessel and make it very difficult to identify.

Bacteria also influenced the identification procedure. The remains of the Titanic *might be the best-known example. Mussels prefer shallower waters for their existence, but bacteria thrive at any depth. The* Titanic *is a prime example of how bacteria can deteriorate metal, eating it down and leaving "rust-sicles" to hang off the edges of the ship. Bacteria played a significant role in breaking down wrecks in Lake Superior, which had less mussel damage than the other lakes, though bacteria also factored into shipwrecks' decline on the other four Great Lakes.*

Ken Merryman and Jerry Eliason were familiar with this, of course; they'd seen it on other wrecks they'd explored. But first they had to find something to identify. They knew from extensive experience to carry on without excessive expectations. This was their first day at this location, and while they had chosen it because their updated research directed them there, they were covering a considerable area. Finding a wreck on the initial day of searching was very uncommon.

The Heyboy *carried on, sailing with promise. The men kept an eye on the monitor. And then, with no accompanying fanfare, it happened: a large object was visible on the lake floor. It was as large as a sunken vessel, but it was sticking upward at a strange angle. The image on the monitor gave only a murky shape, and it disappeared in short order. For all the searchers knew, it could be a large school of fish.*

The men turned the boat around and repositioned themselves over the object. It looked like a shipwreck, but was it the PM 18, *as she was sometimes called? Yes, the angle was unusual, but not unheard of. When they had found the* Manistee *two years earlier, the wreckage had been at an unusual angle, but not quite as steep.*

They lowered Jerry Eliason's camera to the wreck. It was very deep, around five hundred feet, and all they could positively determine was that it was a wreck. Rather than attempt further exploration, they decided to return the next day, when they had more time and, with any luck, less wind.

〰

ON SEPTEMBER 8, 1910, Captain Peter Kilty waited dockside at the harbor in Ludington. The Steamboat Inspection Service was giving the *Pere Marquette 18* its thorough annual examination, and by late afternoon Kilty was growing fidgety about loading his cargo and moving across the lake, bound for Milwaukee. His vessel's sister ship, the *Pere Marquette 17,* had already arrived in Ludington, unloaded, reloaded, and departed for her own trip to Milwaukee. She, too, carried a load of boxcars and miscellaneous freight.

At fifty, Captain Kilty was the eldest of the Marquette fleet's commanders, and he was highly respected by his peers and those who sailed under him. Like his father before him, he was a sailor for life, as comfortable on water as he was on land. He'd spent his first twenty years on Beaver Island, Michigan. He then moved with his family to Onekama, Michigan, and began working on fishing boats and, occasionally, larger schooners. His ascent up the ladder of command was quick: he was given command of a fishing tug at twenty-nine years of age and moved on to captain the *John D. Moore,* a passenger steamer, for a season. He hooked up with the Northern Michigan Line and worked for six seasons on the *Petoskey,* as first mate and, eventually, master.

The *Pere Marquette 18* before disaster

Kilty's easygoing demeanor seemed to defy the gravity of his authority. He was rarely rattled; when the occasion arose and he had to assert his command, he did so with a smooth confidence that provided assurance to officers and crewmen. He had joined the Marquette company in 1893 and had commanded the original *Pere Marquette* and *Pere Marquette 17* prior to being appointed captain of the *Pere Marquette 18.*

Kilty's boat passed the inspection, and the captain ordered her loading. Twenty-nine loaded railroad cars—nearly full capacity—were secured, and passengers began boarding. Kilty was probably relieved by the inspectors' findings. The *Pere Marquette 18,* he felt, required maintenance after the beating she had taken during the summer months. According to the contract terms, the vessel was to be returned in good condition at the end of the tourist season. This had not happened in 1910, to the extent that Kilty worried about the shape his boat was in when he assumed command for fall shipping. "What have they done to my boat?" he reportedly said when he saw her deplorable condition when she was returned after the summer season. Kilty checked the hull plates; some had come loose, so much so that he ran the pumps constantly. He told his wife that it would probably take a miracle for the vessel to cross the lake to Milwaukee. Company officials did not take him seriously when he voiced his concerns.

Kilty overlooked his reservations when he left Ludington at 11:40 and guided the *Pere Marquette 18* west toward Milwaukee. The winds were fresh and the seas were running high. A possible early-fall storm was on the horizon, but the weather conditions were all too familiar to the seasoned skipper. The weather, as it turned out, presented no problem, and the *Pere Marquette 18* sailed without a hitch until she was halfway across Lake Michigan. But at about 4:30 a.m. on September 9, an oiler, on his way to oil some bearings, noticed a large amount of water—seven feet—accumulating in a compartment near the flicker (crew's quarters). The alarmed sailor reported his discovery to the chief engineer, who, in turn, delivered the news to the pilothouse.

No one panicked. The water, the oiler deemed, was entering through an open deadlight (porthole), which was closed without much effort. Captain Kilty, off duty and asleep in his room, was alerted. He examined the situation and likewise displayed little concern. They would proceed to Milwaukee, as planned.

The open deadlight, however, was not the major problem. Water was gushing into the boat from another source. The stern settled deeper in the lake. Three additional deadlights, now underwater, gave way. With the stern flooding and no solution available, Kilty ordered a change of course. Rather than take any unnecessary risks, the vessel would head to Sheboygan, which was closer and would save them a couple of hours of sailing time.

The decision would be debated over the ensuing weeks. The *Pere Marquette 18* was in trouble—that much was certain. With the rudder and propeller deeper in the water than usual due to the stern slipping lower, steering the boat was becoming problematic. Captain Kilty had all the pumps operating, but they couldn't handle all the water entering the vessel. Going back to Michigan wasn't an option; the boat was closer to Wisconsin. Kilty discussed options with his fellow officers. He was remarkably calm, despite the peril. Did he believe he could sail to Sheboygan before the flooding took its inevitable toll? Would the flooding enter the engine room, knock out the boilers, and leave his vessel dead in the water? Should he call for immediate assistance? What was the proper action to take? The answers to such questions would never be known.

The boat may have risen, but it was only temporary relief. Water continued to fill the stern of the boat. Captain Kilty ordered Stephen Sczepanek, the twenty-two-year-old wireless operator, to call for help. Kilty had no choice but to concede that the *Pere Marquette 18* was likely to sink before she reached land. He ordered the distribution of life jackets.

Sczepanek, a Massachusetts native, was employed by the United Wireless Company and was on board the *Pere Marquette 18* in a work-for-hire capacity. This was his second vessel problem that year. He'd been on board another vessel, the *Arizona,* on Lake Michigan in January 1910 and all had ended well; two boats came to the *Arizona*'s rescue and towed her to Chicago with no loss of life or vessel. This occasion was different.

Beginning at 5:00 a.m., Sczepanek sent out a series of dispatches, each more urgent than its predecessor. "Car ferry 18 is sinking midlake—help—answer—answer," went the first message. Twenty minutes later, three transmissions were sent:

> No. 18 is sinking midlake, for God's sake help us.
>
> No. 18 is sinking between Ludington and Milwaukee for God's sake save us.
>
> Help, quick. Carferry 18 is sinking.

A number of boats received the transmissions. The *Pere Marquette 17,* heading back to Ludington following a stop in Port Washington, a port just outside Milwaukee, responded and sailed to the *Pere Marquette 18*'s reported position. It was a couple of hours away, but there was always the chance that the crew might pluck survivors from the water.

Meanwhile, Captain Kilty oversaw the loading of the port lifeboats, each manned by a crew member acting in a supervisory capacity. The crew members were to move the lifeboats away from the *Pere Marquette 18* to avoid being hit by the vessel or being sucked down if and when she sank.

Amazingly, she was still afloat when the *Pere Marquette 17* arrived at 7:15 a.m. The *Pere Marquette 18*'s stern had settled to the point that it was obvious she was going to sink. The *Pere Marquette 17*'s captain, Joseph Russell, wanted to pull alongside and take passengers, but Kilty still hoped to save his boat. Shouting across the water, Kilty asked that the *Pere Marquette 17* follow him at a close distance and come to his aid if it looked like his boat was actually sinking.

All this took place in about twenty minutes. Then, without further warning, the *Pere Marquette 18*'s stern sank very suddenly, send-

A drawing depicts the sinking of the *Pere Marquette 18* on September 9, 1910, with the *Pere Marquette 17* in the distance.

ing the bow high into the air. The smokestacks broke free, one being launched about thirty feet. With a gigantic rush of water, the boat sank.

Wheelsman Simon Burke, a twenty-year veteran of sailing the lakes and survivor of prior shipwrecks, remembered the horrible sinking of the *Pere Marquette 18* and his survival of the incident:

> When I realized that the boat was at last sinking I ran out on the bridge with a life preserver which second mate Walter Brown brought me, and dropping to the main deck I plunged overboard just as the boat disappeared. I must have been in the center of the whirlpool, for I recall having spun around like a top, drawn irresistibly downward, it seemed, 100 feet. Then I rose to the surface and barely had time to catch my breath when I went down again and was under for a long time. After I rose the second time I seized a piece of the cabin and managed to remain afloat until I was picked up.

Captain Kilty went down with the ship, as did all his officers, leaving nothing but speculation about the details of the *Pere Marquette 18*'s demise. The men trapped in the engine room perished as well. Stephen Sczepanek remained at his post, desperately seeking help until it was too late to abandon ship; he became the first wireless operator to die while in service on the Great Lakes.

Passenger Seymore Cochrane clung to a cabin door for an hour and ten minutes prior to being picked up by the *Pere Marquette 17*. The Chicagoan had been reading a magazine in his room when a cabin boy beat on his door and told him that the boat was sinking. As a survivor, Cochrane credited the boat's crew, many of whom perished when the boat went down, for saving him and others. Their actions, he said, were selfless and heroic. "The boat was well equipped with life boats," he told an interviewer, "and if the members of the crew had not been cock full of manliness they might have saved themselves and let the passengers perish. But thank goodness they were not that kind of men."

Early in the morning the day after discovering a sunken vessel that they hoped was the Pere Marquette 18, *the shipwreck hunters returned to the scene to explore the wreck at least by camera. Their hopes ran high. There was no other recorded wreck in the area, and they could think of no ghost ships in the location of their discovery.*

After a long ride back to the area, they found the wreck and dropped a camera near it. As they feared, mussels covered it, ruining their chance of an easy identification. They would have to use their knowledge of ship construction to help them figure out the boat's identity, and it was going to be a challenge.

"It was different from what we'd seen before," Merryman recalled nearly three years later. The pilothouse, for example, was at the stern of the wreck, not in the place it would have been on the Pere Marquette 18. *"That threw us off for a while," said Merryman, who would ultimately learn that the pilothouse had broken away during the sinking and wound up settling where it did when the boat hit the lake floor, stern first.*

They had multiple questions, many of which were answered when they looked at the wreck's midships section. The general construction of the boat, noted Merryman, "matched identically to the Pere Marquette 18."

"There was absolutely no inconsistency," Eliason agreed. If they needed any more proof, he added, they found it when they saw two railroad cars on the lake bottom. They eventually noticed lifeboats nearby and, said Merryman, a particular type of bulkhead sticking out of the rear of the wreck.

THE CREW OF THE *PERE MARQUETTE 17* was no less heroic in their efforts to save survivors and recover the dead. The lifeboats were lowered, two fanning out toward survivors bobbing on the lake's surface. Unfortunately, a third lifeboat—the first launched—met a tragic end. After it was lowered, heavy waves lifted it and smashed it against the hull of the *Pere Marquette 17.* Two of the *Pere Marquette 17*'s mates were killed and two others injured.

Twenty-seven survivors were taken aboard the *Pere Marquette 17,* hoisted to safety by ropes secured to the boat's lifeboat davits. Other vessels arrived later in the morning. Most had heard the desperate SOS calls of the *Pere Marquette 18*, but none had been close enough to be of any assistance. By the time they did arrive, most of the survivors had been pulled from the water. The additional boats were able to pull out six more survivors. Of all those on the *Pere Marquette 18,* only thirty-three lived through their ordeal. A very small percentage of the dead were tallied, and most of those had been dragged down with the boat and released. Among the recovered was Peter Kilty, whose remains were sent to his hometown for a respectful memorial service and burial.

Respect was largely absent when the events leading to the sinking were carefully examined. Theories abounded as to what, exactly, brought the boat down. The vessel's officers—those best equipped for conjecture—had gone down with the ship, and the surviving crew and passengers were lost for an explanation. The easiest explanation, that the *Pere Marquette 18* was in poor repair and not seaworthy when she left Ludington, was rejected by those who had inspected her before she departed for Milwaukee. The idea that the boat had filled with water rushing in through open or destroyed deadlights was bandied about; there was a lot of credence but no absolute proof to this theory. Another theory, advanced by several witnesses, suggested that the boat sank due to water boarding the vessel when the railroad cars were being scuttled.

Captain Kilty was harshly criticized for some of his actions—or lack of action. To hear some tell it, he had been too slow in reacting to many factors that placed his vessel in peril; even at the end, when it was clear that the *Pere Marquette 18* was slowly succumbing to the water weighing her down, Kilty was guilty of trying to save his boat when he should have been finding a way to remove passengers. Kilty, of course, was not around to explain his actions or defend himself.

The Steamboat Inspection Service listened to all the testimony from survivors and miscellaneous experts and reached a safe conclusion. "Investigated foundering," it wrote, "but could come to no definite conclusion as to how the water filled the after compartment, which was full of water at the time of sinking."

No one would ever know the details of what sank the *Pere Marquette 18,* only that twenty-seven people—and possibly more—had lost their lives.

And no one knew, for more than a century, where the boat had come to rest.

Rouse Simmons
(1912)

SCHOONERS RULED THE GREAT LAKES when the *Rouse Simmons* was constructed by Allan, McClelland & Company in 1868. In the days prior to the discovery of the iron deposits in Minnesota and the subsequent ever-growing demand for iron ore, schooners shipped lumber and grain, as well as all types of supplies, throughout the upper Midwest. They would visit port cities and drop off building supplies to the rapidly expanding metropolitan cities. The *Rouse Simmons,* named after a well-known businessman in Kenosha, Wisconsin, was built for such a purpose.

Allan, McClelland & Company was a highly respected shipbuilding firm based in Milwaukee. The *Rouse Simmons* was a 123.5-foot, three-masted schooner—not the longest of boats constructed during that era, but a sturdy wind-driven vessel capable of transporting lumber at a quick speed. This was ideal for her first owner, Charles H. Hackley, a wealthy lumber baron in Muskegon, Michigan, who owned a fleet of Great Lakes vessels. Hackley's boats worked the Michigan coastlines, with some (including the *Rouse Simmons*) venturing down Lake Michigan to Chicago. By 1883, the boat's fifteenth year of service, the *Rouse Simmons* was sailing from Grand Haven to Chicago on a weekly basis.

By this time, schooners, aging from years of hard work, were being replaced by steam-driven boats not dependent on the elements to be effective, and wooden hulls were being replaced by metal capable of taking more punishment and hauling bigger cargoes. The *Rouse Simmons,* though hardly old by lake carrier standards, was becoming obsolete.

She changed owners and captains over the years, each hoping to squeeze some measure of profit out of a boat that was finding the bigger, stronger, and faster freighters to be very tough competition. This, of course, translated into close attention to the costs of overhead and maintenance. The boat still had to pass inspections and remain seaworthy, but upkeep was marginal. Whenever possible,

The *Rouse Simmons*

captains found new ways to earn profits from their vessels. The earnings might be small, but they helped pay the bills.

One way to pick up cash that as many as two dozen vessels participated in was the sale of Christmas trees. In the late nineteenth century, more and more homes were following the European, especially German, practice of installing and decorating a fragrant pine tree in the house. This presented an opportunity for lumber schooners. Each year, when the unpredictable November weather made shipping on the lakes more hazardous, some captains closed out their season with a run to upper Michigan for a load of trees, which they would buy from locals (sometimes Native tribes) and load (or overload) onto their boats. They would then sail down the length of Lake Michigan to Chicago, where they would assemble on the Chicago River and sell the trees to customers eagerly awaiting them. After the expense of the trees and the crewmen working on the boats transporting them, there wasn't a huge profit, but it was worthwhile.

The *Rouse Simmons* at pier

August Schuenemann and his younger brother, Herman, each the captain of his own vessel, partnered in the Christmas tree endeavor. They'd take a boat (usually the one commanded by August), load it with thousands of trees in Michigan, and sail down to Chicago. There, near the Clark Street Bridge, they would string lights to decorate the boat festively and sell trees to families who boarded the boat and handpicked them. Both men enjoyed the process. August became known as "Christmas Tree Schuenemann"; over the years, when he was working on his own, Herman would be known as "Captain Santa."

August had been in the Christmas tree business for about a quarter century when disaster struck. While delivering a load of trees to Chicago in November 1898, his boat, the *S. Thal,* confronted a vicious storm and lost the battle. August and his crew were lost with the vessel. Herman would have perished with his brother, but this year he didn't work in the Christmas tree trade. Herman's wife, Barbara, had recently given birth to twin daughters, Pearl and Hazel, and Herman had decided to stay home with his wife, the babies, and his older daughter, Elsie, instead of moving and selling trees. He was devastated at the loss of his brother, but he, like August, recognized the hazards of sailing Lake Michigan in November.

The tragedy did not deter Herman Schuenemann. He was back at it the following year and the thirteen years that followed. Two years after losing his brother, he lost his boat, the schooner *Mary Collins,* in a freak accident near Little Harbor, Michigan, a small town near Thompson, Schuenemann's Upper Peninsula destination. He mistook a kerosene lamp burning in a second-story window for the Thompson harbor light and directed the *Mary Collins* toward it. She grounded in shallow water, but all those on board survived.

Schuenemann's commitment to the Christmas tree business was probably impossible for others to understand; his kind, sparkling blue eyes, bushy mustache, and friendly smile made him an engaging personality that might have reminded strangers of a favorite uncle. He was known for his generosity. The average tree cost less than a dollar, but not all families could afford that amount. Schuenemann

Captain Herman Schuenemann *(center)*, 1909

was well known for giving away trees to the disadvantaged and to churches; he kept, in his wallet, news clippings about his generous nature. More than one account commented that the holiday season wouldn't exist in Chicago without Captain Santa.

People might not have known that he paid for every tree he gave away, and he needed every penny he could earn from his paying customers. His entire life, dating back to boyhood, had been hardscrabble, and now, with a family to support, he felt the effects of an endless race against poverty. He tried to find ways to get ahead. He bought a tavern in Chicago. He squirreled away enough money to purchase land—with pine trees—in Michigan.

In 1910, facing bankruptcy, he tried to bail himself out of trouble by buying one-eighth ownership of the *Rouse Simmons*. He carried on as always: commanding a boat and delivering lumber during the shipping season, and visiting the Chicago docks with trees on his final trip of the season.

It all came to an end in 1912.

The *Rouse Simmons* with trees

THE WIND WAS FRESHENING when Herman Schuenemann and his crew began loading the *Rouse Simmons* for her annual Christmas tree run. The weather was never easy to predict in the days before radar, jet stream observations, GPS, and other measures taken for granted today. Schuenemann trusted his instincts and experience. There might be a little rough sailing over the next few days, but it was late in the season, and there was no telling what might be coming on the heels of what appeared to be a minor disturbance.

Loading the boat was a time-consuming procedure. The small and medium-sized trees were stacked high in the vessel's hold. The larger ones were lashed to the deck. The sweet smell of newly cut evergreens filled the air. The boat settled lower in the water as between 3,500 and 5,000 trees occupied every square inch of available space. Schuenemann had been through this so many times that all that really mattered was getting as many trees loaded in as little time as possible. He was eager to get the trip under way.

It's not known how many men were on board the *Rouse Simmons* when she set sail onto the lake in the waning hours of daylight for the final time, but it's estimated that sixteen to twenty-three took their places on the doomed vessel. This number included several lumbermen who hoped to reach Chicago in time for Thanksgiving festivities. The ever-accommodating captain invited them aboard.

The stormy weather intensified. The barometer fell and air temperatures dropped. The wind increased significantly. In the *Rouse Simmons*'s younger days, when she was one of the bigger schooners among the many plying their trade in what has become known as the golden age of schooners, the gathering storm might have been more annoyance than threat. We will never know for certain, but it's probable that the rickety ship was taking a beating by the time she reached midlake. If Herman Schuenemann had gambled that he might run ahead of the storm, he was losing his bet.

There would later be disagreement (as often was the case with storms that took down boats, with loss of life) about the severity of the storm, when the worst of it occurred, and where it took place. In remembering the storm in 1990, Fred Neuschel, writing for *Anchor News,* the official publication of the Wisconsin Maritime Museum, claimed that "the gale of November 23rd proved itself to be the most deadly storm of 1912." Other publications, while noting that the storm had been particularly nasty, placed the worst of it in early December rather than a week or so earlier.

Then there was the matter of localization. A storm might be fierce in parts of the lake while relatively calm elsewhere. Captain Bernie Cooper, a weather authority who was skipper on the *Arthur M. Anderson,* the vessel following the *Edmund Fitzgerald* in the storm that brought the *Fitz* down in 1975, spoke of the ways location affected sailing on the lake. He'd been in worse storms during his career, he said, but never in one that was as terrifying in certain locations as the one that sank the *Edmund Fitzgerald.*

Was Herman Schuenemann fighting this phenomenon when he took the *Rouse Simmons* onto Lake Michigan on November 22–23?

HOW AND WHY the *Rouse Simmons* sank cannot be known for certain. She was, quite simply, a ghost ship—a vessel that left harbor never to be heard from again. The location of her wreckage remained unknown for nearly six decades, and when it was found, the discovery was a sort of happy accident: a diver searching for another vessel found her instead.

The mystery brought legend and myth in her wake. Sailors, a superstitious group by nature, told wild tales of seeing her ghostly outline, or of hearing her bell, in the area where she was believed to have sunk. A number of accounts of the sinking existed, some far more dramatic than others. The more colorful descriptions made a point of mentioning that rats were seen abandoning the boat, or that one or more of the crewmen refused to sail before the *Rouse Simmons* left Thompson, supposedly believing that the boat was not seaworthy. Given the vessel's age and history, that would not have been difficult to believe.

A nasty storm featuring high winds, heavy seas and snow, and plunging temperatures assaulted the *Rouse Simmons* as she made her way down the lake. The moisture soaked the trees piled high on the deck, the cold air froze the moisture, and the *Rouse Simmons*—already overloaded—took on greater weight, further reducing her freeboard in the stormy seas. A very concerned Captain Scheunemann sent two crewmen to check the lashings on the deck, but both were swept overboard by waves boarding the boat. Water, whether from those waves or leaks in the boat's hull, flooded the *Rouse Simmons.* In the predawn blackness on the lake, Scheunemann might have thought about his brother and his fate. His vessel appeared to be sinking, and he had little recourse to save her.

The *Rouse Simmons* persevered. As the story went, a spotter at the Kewaunee Life-Saving Station saw a schooner (he couldn't have known her identity from its distance in the storm) at 2:50 p.m. The vessel was struggling, her sails were tattered, and her flag was flying at half-mast—a universal signal of distress. The battered boat was being pushed south by the gale-force wind.

The surfman told his commanding officer, Captain Nelson Craite, what he had seen. "I immediately took the Glasses and made out that there was a distress signal," Craite recorded in his logbook. "The schooner was between 5 and 6 miles E.S.E. and blowing a Gale from the N.W." He issued an order for the station's gas-powered tugboat to check it out, without any luck. The boat was already out. Craite called the Two Rivers Life-Saving Station and explained what he had seen to George E. Sogge, the station keeper. The schooner, Craite said, was moving at a speed that should take her near Two Rivers at any time.

Sogge dispatched a powerboat to look for her. Those on board the powerboat, the *Tuscarora,* scanned the horizon in every direction, endangering their own safety to assist a vessel in trouble. They saw nothing. The schooner was nowhere on the surface of Lake Michigan.

Another possible explanation for the loss of the *Rouse Simmons* contained equal mystery. The boat was not driven down by a storm; her sinking, instead, was the result of age and conditioning. According to weather observers, the storm that supposedly snow-blinded the boat did not kick in until five in the afternoon on November 23. The seas were running high, but a seaworthy vessel, commanded by an experienced skipper like Herman Scheunemann, could have easily functioned. At least two captains reported seeing the *Rouse Simmons* in the Kewaunee area, and neither believed the vessel to be in any kind of trouble. Her flag, they said, was not at half-mast, or else both boats would have gone to the assistance of the stricken vessel. The noteworthy conflict between their reports and that of the Kewaunee station only increased the air of mystery surrounding the boat's disappearance.

Speculation swirled around the vessel's condition. Charles Nelson, a minority owner of the *Rouse Simmons* and on board during the final voyage, was said to have been so concerned about its condition that he seriously considered canceling his plans to sail on her. His young daughter cried and begged him not to go. In the wake of the disaster, sailors, including crewmen, spoke pointedly about the boat's poor condition. An aging schooner with a questionable reputation had no business being on the lake.

The news media contributed heavily to the confusion. Always in a rush to scoop their competition, newspapers ran headlines that were either untrue or only partially true. CHRISTMAS TREE SCHOONER SIGHTED / SANTA CLAUS SHIP MAY BE SAFE, a *Chicago American* headline shouted on December 5, nearly a week after the disaster. The same day, the same paper reported MORE WRECKAGE FROM THE CHRISTMAS SCHOONER *ROUSE SIMMONS* WASHES ASHORE. In Milwaukee, the morning daily's headline proclaimed CHRISTMAS SHIP SAFE, SAY MARINERS. The only thing that remained consistent throughout the ordeal was the unrelenting interest in the story, made more compelling by the Christmas tree angle.

The shipping community held out hope for the *Rouse Simmons* for days after she missed her scheduled arrival in Chicago. The storm on the lake had been brutal enough to give people hope that the Christmas Tree Ship had taken shelter somewhere and had yet to make an appearance in the Windy City. This was common practice. A boat would seek cover in the lee of a storm, or anchor in a harbor, until the storm passed and it was safe to sail again.

One could not blame those who might have otherwise seemed unreasonable in holding on to hope. The storm had gained in its intensity on November 23. Winds out of the northwest blew at gale force. Heavy snow fell, decreasing visibility. Captains and other sailors out on the lake swore the storm was one of the worst they had experienced in years. It would have been completely understandable if a skipper ordered his wheelsman to duck out of trouble.

The sad truth about the *Rouse Simmons*'s demise was discovered in the weeks and months that followed. First, Christmas trees washed onto the beach near Two Rivers. This, by itself, was not proof of sinking; it was quite possible that Schuenemann, in an attempt to lighten his vessel, had ordered them to be cast overboard. Or the trees simply could have been washed overboard by heavy waves. No one knew.

More definitive answers came later. Not only did more trees end up on the eastern Wisconsin shore, but fishermen found them caught in their nets. On December 13, a bottle was picked out of the water by a fisherman near Sheboygan. A note, written on a logbook page, read:

> Friday. Everybody goodbye. I guess we are all through. Sea washed over the deck on Thursday. During the night the small boat was washed over. Leaking bad. Ingvold and Steve overboard Thursday God help us.

The note was signed by Herman Schuenemann.

Still, the mystery surrounding the lost schooner never let up. Until the wreckage of the boat was located, the *Rouse Simmons* was a ghost ship with an interesting past.

The loss of the *Rouse Simmons* did not end the Schuenemann tradition of selling Christmas trees near the Clark Street Bridge. Herman's wife, Barbara, and their daughters continued the tradition until Barbara's death in 1933. The trees did not arrive by schooner or any other type of lake vessel, however. They were shipped to Chicago by rail and transferred to a boat anchored (and decorated) near the bridge. People were still invited to choose their trees on board, and the Schuenemanns provided refreshments for all. This became a way of keeping Herman Schuenemann's spirit alive.

In 1924, nearly a dozen years after the sinking of the *Rouse Simmons,* a fisherman found a major artifact. Tangled in his net was a small package, purposely made waterproof by oilskin wrapping. Inside was a wallet belonging to Herman Schuenemann. It still contained the press clippings about the Christmas Tree Ship and her captain. The wallet was returned to his family.

~~~

KENT BELLRICHARD, a Milwaukee sport diver and shipwreck sleuth, discovered the wreckage of the *Rouse Simmons* on October 30, 1971, while searching for another boat, the *Vernon,* a 177-foot vessel that
~~~

sank in a storm in October 1887. Bellrichard's sonar indicated the presence of a large object in 172 feet of water. Although the weather was overcast, threatening a possible storm, he dove down to the site, but his hand-built lighting system failed as he was reaching the boat. Not willing to return to the surface until he had at least touched the wreck, Bellrichard probed in the darkness until he was feeling his way along the hull. He was fairly certain he had found the *Rouse Simmons,* but he needed much more proof.

"It was a tremendous thrill the very first time I went down, because I was 99 percent sure it was the *Rouse Simmons,*" Bellrichard said in 1975 during a television interview. "In fact, I came back up and got in the boat, and there wasn't anybody within six miles. It was quite rough on the lake . . . and yet I was hollering to myself in joy that I had finally found this ship."

He had to wait for another opportunity to explore his finding. The weather was uncooperative, but he was finally able to make another dive a week later to gather conclusive evidence. When he touched the trees still lashed to the deck of the wreck, he knew he had found one of the most fabled of Great Lakes ghost ships. With good lighting, he could read the name painted on the wreck, confirming that it was indeed the *Rouse Simmons.*

The vessel had come to rest upright on the lake floor. The trees belowdecks were stacked as they had been nearly six decades earlier, the needles on many of them mostly intact. The boat's three masts were detached from the vessel but were located nearby. An anchor lay on the lake floor, well in front of the boat, its chain still attached to the bow. Perhaps most disturbing, the *Rouse Simmons*'s wheel was missing and could not be found anywhere near the wreckage.

What did this suggest about the Christmas Tree Ship's last moments? One theory, popular long before Bellrichard's discovery, made sense: The weight of the trees was enormous, and those on deck were covered with ice. Cargo was known to shift in heavy seas, and if this happened with the *Rouse Simmons,* overloaded as she was and with reduced freeboard, the weight might have shifted forward and driven the boat, bow first, in a nosedive to the bottom of the lake.

Kent Bellrichard *(left)* and two colleagues with the *Rouse Simmons* artifacts, 1971

Another theory focused on the anchor. The *Rouse Simmons* could have lost her wheel in any number of ways, but one thing is certain: without the ability to steer, she was at the pitiless mercy of the storm. Huge waves pushed her around at will. The *Rouse Simmons* was unable to head toward safety or be directed in a way that avoided her being caught in troughs between waves. She was doomed. Recognizing his precarious position, Herman Schuenemann might have ordered the lowering of an anchor as a last resort. The wind, out of the northwest, would have been pushing the *Rouse Simmons* away from the lowered anchor, which explained its position to the wreckage.

These and other ideas could not be proven, nor did it matter. They might make for fascinating discussions and disagreements in academic institutions and in bars. Artifacts removed from the *Rouse Simmons* (a bottle, a wooden stool, a piece of tinder from the vessel, and other bits) were placed in the Rogers Street Fishing Village museum in Two Rivers. The anchor was brought up and placed near the entrance of the Milwaukee Yacht Club. Trees were removed and distributed to institutions in Wisconsin and Illinois. In 1999, a fisherman snagged the wheel a mile and a half from the wreckage; it was carefully restored and placed alongside other artifacts in the Two Rivers museum.

Time was kind to the *Rouse Simmons.* Her story was repeated, usually at Christmastime, in newspapers and magazines. Songs about her were written. A play was produced and documentaries were filmed. There appeared to be no end to a story that had begun simply and ended tragically.

Eastland
(1915)

PEOPLE BEGAN GATHERING dockside at the Chicago River as soon as the break of daylight. They came from all over the area, many knowing little or no English, almost all members of the working class, folks of all shapes and sizes, eager to enjoy the big company picnic. It was cloudy, with mist and rain in the air, but this didn't dampen their spirits. This was *their* day.

The women wore decorative bonnets and print dresses, the men bowlers, straw hats, and suits. The men carried baskets containing extra clothing, swimsuits, or food. Their children, dressed in festive clothing, stuck close to their parents, not wanting to get separated and lost in what was turning out to be a large crowd.

It was July 24, 1915, the day of the Western Electric Company's annual picnic. Those attending had paid a dollar per adult and fifty cents per child to take one of the six boats presently tied to the docks in Chicago to Michigan City, Indiana, a pleasant trip of forty miles. They would spend a day in revelry, singing and dancing and swimming and picnicking. For the younger men and women, there was the opportunity for chance meetings and maybe (who could say?) romance.

The *Eastland* in Cleveland, Ohio

Throngs of people headed toward the *Eastland,* the first boat scheduled to depart and by far the biggest and most attractive of the forms of transport. The *Eastland* was a fantastic part of the Chicago maritime landscape, a sleek 269-foot passenger vessel capable of slicing through the water at twenty-two miles per hour, earning her the nickname Speed Queen of the Lakes.

Although those clamoring to head up the plank to the boat had no knowledge of it, the *Eastland* had an up-and-down history on the lakes, a background that might have given many pause had they known.

But this was *their* day, one designed to help them forget about the long hours of hard work that had built Western Electric into an industry megapower and paid their bills, one designed for a feeling—at least for one day—of freedom.

~

CAPTAIN HARRY PEDERSON, the *Eastland*'s skipper, always seemed to have questions about his boat, and this day was no different. It was going to be a busy day, starting with his taking the Western Electric Company's passengers to Michigan City, then ferrying a load of pas-

The *Eastland* before tragedy

sengers from St. Joseph, Michigan, to Chicago, and, finally, back to Michigan City to bring Western Electric's passengers home. He could now hear many were hauling up the gangplanks of his boat, which was good—but he had other measures to consider.

First was the matter of the weather. There was just enough fog to slow him down. Michigan City was not a usual *Eastland* stop, and Pederson didn't need anything to hinder him. He had consulted briefly with Joseph Erickson, the boat's chief engineer, who assured him there was no reason to worry. They would load up and take off without a problem.

Then there was the number of passengers he would be boarding. The *Eastland*'s history was, at the very least, spotty. The actual numbers of passengers permitted on board had varied over the years, depending on the vessel's operations and, more recently, the wreck of the *Titanic*. Though that ship had been a saltwater vessel and sank under extraordinary circumstances, the incident had a huge effect on all waters in the United States. Laws were passed that demanded enough lifeboats and other rescue vessels be added to account for every passenger on a ship. This affected all passenger liners—and some left the business as a result—but, arguably, the *Eastland* was among those hit the hardest. Incidents in her past had resulted in her number of allowed passengers to be decreased.

At her launching in 1903, she was undoubtedly the most beautiful boat owned by the Michigan Steamship Company. The Jenks Shipbuilding Company had supplied her with four decks and the capability to steam at a speed that matched her beauty. Unfortunately, her stability came into question almost immediately: she ran into a tug, the *George W. Gardner*, the year she was christened, costing her captain his job. The following year, she listed heavily in the open water, leading to an investigation that lowered her capacity from 3,399 to 2,500 passengers.

The problems kept building. Sold in quick succession to the Chicago–South Haven Line and the Lake Shore Navigation Company, the *Eastland* moved from Lake Michigan to Lake Erie. Her construction changed as well. Decks were changed, the twin stacks shortened. The newly designed decks offered passengers more room to congregate but did not quell the suspicion that the boat's metacentric height might be incorrect—to such an extent that in 1910, the Eastland Navigation Company, another new owner, took out a full-page newspaper advertisement offering a five thousand dollar reward to anyone who could prove the *Eastland* unseaworthy. The challenge brought no takers.

The *Titanic* disaster of 1912 changed everything. The new obsession with passenger safety—long overdue—spelled additions and new potential problems for the *Eastland.* Three lifeboats and six life rafts were added, and tons of concrete were brought on to strengthen decks weakened from age. These modifications further altered, negatively, the vessel's metacentric measurement. The *Eastland* was a vessel on a crash course with fate.

Captain Pederson, who had been with the boat since it was sold to the St. Joseph–Chicago Steamship Company, was aware of potential problems, as he knew about some of the improvements the company had built into the ship. Those in charge had added more life jackets, and their efforts had been rewarded, after another inspection, with an increase to 2,500 passengers.

Now, with passengers piling on and taking places wherever available, Pederson had to trust the instincts of his chief engineer, who happened to be the son of one of the boat's recent inspectors. Ballast was added to steady the boat, but it only seemed to help for a few minutes before another change was demanded.

In a formal statement offered to the investigators shortly after the accident, Harry Pederson still was confused as to what exactly had happened:

> I was on the bridge and was about ready to pull out when I noticed the boat begin to list. I shouted orders to open the inside doors nearest the dock and give people a chance to get out. The boat continued to roll and shortly afterwards the hawsers broke and the steamer turned over on its side and was drifting toward the middle of the river. When she went over I jumped and held onto the upper side. It happened in two minutes. The cause is a mystery to me.

~

IF THOSE BOARDING the *Eastland* were worried about its stability problems, they didn't show it. Boarding began at 6:30 a.m. People hurried up the gangway to packed positions, usually on the level of the boat or the promenade deck near the stern of the vessel. Young couples laughed about the boat's instability, creating a game out of the way the boat rolled with the influx of new ballast water. Roughly fifty people boarded per minute.

Elsewhere, the crew took it all in stride. In the engine room, Joseph Erickson, although concerned with the boat's tilting, was not worried. The vessel had her particulars, but he was now accustomed to them. He moved about the huge room barking orders to those preparing the engine for launch and those adding to the ballast tanks. Everything appeared in working order, even if he might have planned differently.

In his book *Ashes Under Water: The SS Eastland and the Shipwreck That Shook America,* author Michael McCarthy describes the engine room routine:

> In the *Eastland* engine room, Erickson noticed that his inclinometer, a little metal arrow that swiveled as the ship tilted, was leaning toward the deckside. Unconcerned, he ordered his men to open the valves on the river side, the port one, to correct the ship's slight starboard leaning. "Boys, steady her up a little," Erickson said.
>
> In about three minutes, the *Eastland* was straightened up and the gangplank was level for the parade of passengers boarding. All was back to normal.

The water in the ballast tanks said otherwise. The first adjustment went well, with the boat sitting level shortly after water was added to one tank and eliminated from the other, but the relief lasted only minutes. The *Eastland* began a list in the other direction, straining the ropes holding her to the docks. Erickson ordered another adjustment—yet, as he quickly noticed, each new adjustment led to bigger problems.

Each movement affected materials on board. Bottles went flying, crashing on the deck and turning areas into a sharp mess. One heavy roll sent an unoccupied piano rolling and would-be dancers scrambling to get out of the way. A refrigerator readied for an expected onslaught of customers slid briefly before tipping over, injuring several women and spilling its contents on the floor. People scrambled everywhere.

The boat's crew, now aware of the vessel's impending fate, exited as well as they could, adding to the panic. The *Eastland* tugged at the few lines that held her, snapping them as the boat rolled toward the river. Seeing that the situation was hopeless, Joseph Erickson abandoned his post. Captain Pederson, outside and surveying the boat's potential from the beginning, raced toward the inner dock and held on. His vessel was doomed—and several thousand people were at risk.

To those watching from shore, the catastrophe was quick and final: in only minutes, the *Eastland* rolled over.

Chicago harbor master Adam Weckler recalled the boat's final moments as if it took place in slow motion:

> The boat kept turning and [Captain Pederson] shouted to the people to get off the best way they could, and the boat it should say around in 8 or 10 minutes' time laid right on the side. Of course, the people—passengers—onboard were scrambling to get ashore. Those on the hurricane deck jumped overboard.

In *Eastland: Legacy of the Titanic,* a lavishly documented account of the accident, historian George W. Hilton notes that the boat toppled on its side and settled into the mud between 7:28 and 7:30 a.m., less than an hour after it had formally begun loading passengers. The crashing piano and refrigerator had alerted those on the passenger deck that something very serious was occurring, but those belowdecks had clogged the ways off the boat, blocking exits and inviting

The *Eastland* immediately after rolling over

further panic. People jumped into the water or took their chances on the deck. In the moments just before and after the boat took its fatal roll, bedlam ensued.

Bodies hurtled through the air, either by choice or by the *Eastland*'s movements. The lucky ones were able to swim out of harm's way or, better yet, cling to the overturned boat and crawl aboard. Many, especially children, were unable to swim; they drowned as soon as they hit the water. Those trapped belowdecks stood no chance. The boat's rolling would have hurled them hopelessly, the water boarding the vessel in the minutes before her rolling quickly ended many of their lives.

The *Eastland* came to rest in deep but still relatively shallow water, her bow in about twenty feet of water and her stern in roughly three feet. For many of the more than 2,500 aboard, the *Eastland*'s final moments afloat were their final moments. People screamed and shouted as she rolled.

Members of the crew escaped harm free, for the most part. All recognized the trouble at hand, and almost all were able to get off the boat before she took her roll. Captain Pederson hung on to the bitter end, trying to manage an escape route for passengers, but according to Algernon Richey, who watched the accident from land, he managed to escape easily. "[He] stood on the right-hand side of the bridge, with his hand on the rail," Richey would remember. "As she went over, he grabbed it with his left hand and climbed over, and never even got his feet wet."

Chief Engineer Erickson stayed in the huge engine room as long as possible after it became clear that his attempts to steady the *Eastland* were not working and that the boat would probably head onto her side. Erickson had grown concerned about the boilers, which were hot and likely to explode when hit with the river water now flooding in. An explosion, with this many people on board, was certain to be catastrophic. Erickson somehow reached the boilers and turned on the injectors in an effort to cool them before water hit. Then he looked for a way off the *Eastland*. That he survived is miraculous—a true

An internal view of the overturned *Eastland*

testament to good fortune combined with his knowledge of the vessel's makeup. Water was quickly rising as Erickson found his way onto an air duct. He made it to a porthole and was struggling to get through when watchman Robert Brooks found him and helped him out. He entered a chaotic world, but he had escaped certain death.

Rescue efforts had already begun. The *Kenosha,* a tugboat attached to the *Eastland* to guide her down the river and onto Lake Michigan, was not affected by the rollover; she had been cut loose by an ax before the *Eastland* rolled. The tug immediately swung into action, plucking as many passengers from the water as she could hold. The *Kenosha* then found herself acting as a bridge between the stricken boat and the shore.

There were many heroes on this day when so many perished. A nearby fire station immediately sent men with supplies and tools. A factory warehouse acted as a temporary morgue, housing scores of victims until their families could arrive and identify them. Hospitals and the Red Cross, quickly learning of the disaster, sent personnel and equipment. Individuals on dry land lent whatever services they

Eastland recovery efforts

could provide; some threw anything floatable into the water, while others helped survivors to safety.

The river's soft current carried some of the victims away. With so many people in the water, motorized vessels had to shut down. Many of those on other boats, seeing the drama unfold and knowing people on the *Eastland,* rendered assistance while those still trapped beneath the *Eastland*'s decks screamed and pounded for their lives.

Pandemonium was the day's only consistency.

~

THERE WAS NO WAY of accurately assessing how many lived and how many perished in the accident. There were the counted, of course, but there were the uncounted as well. The official tally of passengers was not to be trusted; children were treated differently from adults, some counting as half an adult, some not counted at all. Some passengers were whisked off to hospitals, treated and released, and then taken away by family members. Estimates by those on board and those watching dockside added to the confusion. The range of numbers was staggering. Since final figures on those boarding the *Eastland* were impossible to trust, all kinds of numbers, from modest to preposterous, were given.

C. B. Bradley, a survivor, supplied a great example of how the excitement of the day's rapidly evolving events could mislead interpretations. "People were packed aboard too closely to budge," he stated. "We found the companionways between stairway and promenade so crowded we had to squeeze through. From my observations I should say there were about 3,500 people aboard. Every part of the boat was crowded to standing capacity."

Over the ensuing days, Chicago newspaper readers would receive a veritable smorgasbord of estimations. Some reporters, rushing to provide figures for the next morning's editions of their papers, did their interviews and then the math necessary to come up with believable figures. Others just let the numbers fly, trying to add up those already in the morgue, those trapped on the boat, and estimate how many were living. It was an impossible chore. The *Chicago Tribune,* the city's most read newspaper, offered differing estimates in its next day's Sunday editions. 919 BODIES RECOVERED, one headline shouted. TOTAL ESTIMATED VICTIMS MAY REACH 1299!

As Dwight Boyer, author of *True Tales of the Great Lakes,* explains, the discrepancies might have been unavoidable. Duplication was the norm between the docks outside the *Eastland*; the Reid, Murdoch & Company warehouse held so many of the dead, and the *Theodore Roosevelt* was the vessel handling so many passengers. Reporters weren't clear on how to handle these numbers. The warehouse, for example, knew how many bodies had been recovered so far, but God only knew how long it would take to remove all the dead from the vessel. Were these numbers among those cited by others? And survivors seemed to have different, often self-serving, accounts of the disaster.

The photographs spoke volumes—and many were taken. Practices were different from what they are today: newspapers could—and did—print images of those who perished in the disaster, such as perhaps the lifeless body of a young boy, who hangs listlessly from a man's arms: the man's shocked expression says more about the disaster than any reporter could describe.

These photographs added to the growing portrait of the event: the *Eastland,* on her side in the river, moments after turning over; workers cutting into the boat's hull after an argument with the boat's captain, who felt that they were cutting into the coal deposit; bodies, stacked like cordwood on the docks, covered with sheets and awaiting transfer to the temporary morgue; boats on the river offering assistance to those struggling to stay afloat; relatives in the morgue, searching for loved ones, far too many succeeding in finding the victims they were looking for.

What the photographs could not depict was the sheer horror of a day, begun with such hope and joy and then turned into such hopelessness in little more than an hour. So many people, dressed for festivities and looking forward to fun and games, had lost their lives.

An estimated 844 perished on the *Eastland,* making it the greatest tragedy in Great Lakes history and the third-largest marine disaster in American history, ranking behind only the *Sultana,* which lost 1,653 in a boiler explosion near Memphis on April 22, 1865, and the *General Slocum,* which lost 957 in a fire in New York City on June 15, 1904.

To put the *Eastland* disaster in international terms, the passenger loss was greater than the 829 lost on the *Titanic.* (The *Titanic* had 694 of its crew perish after the liner hit an iceberg in 1912; the *Eastland* lost only three of her crew.) This is not to compare: a single person lost in a shipwreck is a tragedy. It does, however, create a perspective, especially since the *Eastland* sank less than three months after the *Lusitania,* which was torpedoed and sunk on May 7, 1915, at a cost of 413 crewmen and 785 passengers—making the year 1915 one of the worst and most severe in maritime history.

Who was to blame? There would be legal procedures to make that determination.

THE LOSS TO SOME FAMILIES, the world learned, was substantial. Some stood out more than others. The Plamondons ranked highly. Mr. and Mrs. Plamondon had died aboard the *Lusitania,* and now eight more Plamondons had been on the *Eastland.* Seven survived the ordeal; the eighth perished.

One unidentified young boy captured the public's imagination as time passed, his lost and seemingly forgotten body in a makeshift morgue, unclaimed. An increasingly inflamed public demanded some form of justice. What, they wondered, would become of the body known only as 396?

"Who is little No. 396?" the papers wondered. "Who is the 'little feller'?"

No one could say. His parents had apparently died, and there were too many unclaimed bodies to be able to make an identification.

The answer came unexpectedly when two brothers, friends and neighbors of the unclaimed boy, saw his remains and told their uncles that he was Willie Novotny. When the uncles disagreed, the boys insisted that his grandmother be notified. The old woman, who spoke Czech, had already buried her daughter, her son-in-law, and one of their children. For reasons no one would ever understand, she had said nothing about the missing boy. She was brought to the morgue five days after the disaster. She claimed him immediately.

The story didn't stop there. Thirteen thousand Chicagoans showed up to bury him, prompting Mayor "Big Bill" Thompson to declare, "The hearts of all Chicago go out to the sufferers of this calamity. The city mourns."

AND WHAT HAPPENED to the *Eastland*? The boat, full of holes, on her side in the river, was slowly emptied of victims. Law required the owners to move the vessel within thirty days. The Great Lakes Towing Company, run by Will Reid of Port Huron, Michigan, received the job and Captain Alexander Cunning ran the operation with his steamer, a 181-foot boat equipped for a big job. It was ordered that the overturned boat be moved as quickly as possible, as the *Eastland* took up a great chunk of the river.

The boat was raised, but instead of being headed for scrap or being delivered to a yard for repair, as would have been expected, she was sold to the government. World War I was commencing, and the State of Illinois needed boats. Captain Edward Evers hoped to convert the *Eastland* into a training vessel for the Illinois Naval Reserve, and he picked her up for $46,000 in December 1915. Getting the boat paid for and converted required time and money, and Evers did not have enough of either to complete his planned project. The *Eastland* was sold to the navy for $175,000, rebuilt, and renamed the *Wilmette*.

The boat enjoyed a more successful life as the *Wilmette* than she had as the *Eastland*. She operated first as a gunboat and then, eventually, a training vessel, a function she served through World War II. After the war, she was decommissioned and her superstructure was removed; in 1947, she was sold for scrap. The boat was gone, and so too, apparently, was the memory of the biggest disaster in Great

The crowd after the *Eastland* was righted

Lakes history. A spot on the river was dedicated to the Chicago Fire, but nothing existed in memory of the *Eastland.*

Yet the boat never entirely slipped from the city's consciousness, and she was memorialized on significant anniversaries. In 1998 an aluminum plaque was set on the site on the river. It commemorates the event simply:

> While still partially tied to its dock at the river's edge, the excursion steamer *Eastland* rolled over on the morning of July 24, 1915. The result was one of the worst maritime disasters in American history. More than eight hundred people lost their lives within a few feet of the shore. The *Eastland* was filled to overflowing with picnic bound Electric Company employees and their families when the tragedy occurred. Investigations following the disaster raised questions about the ship's seaworthiness and Great Lakes steamers in general.

Books about the *Eastland* would be written, and at least one play, but the boat disappeared, for all intents and purposes, with the big stories rising out of World War II.

Milwaukee (1929)

FROM THE EARLIEST DAYS of Great Lakes shipping, there had been an ongoing (and occasionally heated) discussion about sailing in adverse conditions. The lakes were especially famous for their storms in late fall, but they could be a threat at almost any time. Most companies shut down business during the winter, when portions if not all of the lakes would freeze over, creating hazards that no sailor wanted to face. Some, however, conducted year-round shipping, their large, powerful boats taking on elements with seemingly no worries. Some big freighters even acted as icebreakers, carving out paths for others to sail through.

Captains were divided on how to address the miscellaneous conditions. They were the ones who ultimately had the final say on whether a boat left port in stormy weather—or so went the belief. Critics argued that captains, for sake of job security, succumbed to pressures applied by the companies that employed them. This argument was especially prevalent whenever there was a loss of lives in a shipwreck. The companies would insist that there had been no pressure, that the decision to sail had been made by experienced commanders in the pilothouse, and the captains almost always accepted responsibility—if, in fact, they survived the sinking.

As one might imagine, captains had reputations in the shipping business. Some were known to be cautious, while others were recognized as being willing to sail in almost any conditions. The latter, rarely questioned but occasionally spoken about in less than favorable terms, were given free rein to operate. They were held accountable, by shipping and formal authorities, only if something happened.

One of the more daring commanders on the lakes was Robert McKay, a captain from Scotland who had studied under two of that country's most celebrated mariners. He worked his way up to the position of first mate, sailing in all conditions. After moving to the United States, he earned his captain's certification and built a career on the Great Lakes, gaining a reputation as a no-nonsense presence in the pilothouse and an easygoing, cigar-chomping character outside the work environment. He earned the nickname "Heavy Weather McKay."

McKay proved his sailing prowess and toughness during his five-decade career. While serving as first mate on the *Naomi,* he distinguished himself when the vessel caught fire, and crew and passengers faced the possibility of the freighter burning down to the waterline in the middle of Lake Michigan. McKay calmly supervised the loading of lifeboats, extinguishing a panic that might have led to great loss of life.

With 1929 drawing to a close, McKay figured he would put his lengthy career behind him and retire. The national economy was slipping (the market would crash on October 29), and the future of Great Lakes shipping was uncertain. For two years, McKay had commanded a train car ferry, the *Milwaukee,* which sailed twelve months a year. He was absent from home much longer than he liked. The sixty-seven-year-old announced his intention of retiring at year's end.

~

THE *MILWAUKEE* BEGAN her twenty-seven-year stretch on the Great Lakes as the *Manistique–Marquette & Northern No. 1.* The name took

The *Milwaukee* anchored

no effort to understand: her home port was Manistique, Michigan, and her owner was the Manistique–Marquette Railroad Company. Built by the American Ship Building Company of Cleveland, the vessel was typical of the long, broad, steel-reinforced-hull car ferries of the time. Capable of hauling thirty loaded railroad cars, she moved freely across Lake Michigan, carrying freight between Michigan and Wisconsin and gaining the reputation of being one of the hardest-working vessels in the business. Crossing the lake saved her railroad owners a substantial sum they might have spent in transporting by traditional rail.

The *Manistique–Marquette & Northern No. 1* was constructed at about the same time, and by the same company, as the ill-fated *Pere Marquette 18.* The *Pere Marquette 18* was the shipyard's hull 412; the *Manistique* was 413. The two vessels were similar in design and served similar functions. The *Manistique* was 388 feet long, with a 56-foot beam and a hull of reinforced steel.

In 1909, the car ferry was sold to the Grand Trunk Ferry Company of Milwaukee and had her name changed to *Milwaukee.* Her hull was repainted white for a few years, but very little else about her, including her customary cross-lake route, was altered. She continued to carry an average of twenty-eight cars and did so without noteworthy event until her demise. A year prior to being sold to the Grand Trunk Ferry Company, she was involved in an accident that might have been responsible, at least to some degree, for her sale. On January 5, 1908, she was seriously damaged by ice outside Manistique. She found her way back to land but sank. She received extensive repairs and was put back to work. For nearly the next two decades, she sailed without any problems other than the usual scrapes.

Did the *Milwaukee*'s track record lead to a false sense of security or, worse yet, a sense of invincibility among those who served on her? Some seemed to think so, especially in the wake of her 1929 sinking. She had battled ice, huge seas, heavy winds, and other hazards, and company officials found that the boat was near enough to safety on her trips across Lake Michigan that she wasn't equipped with a radio for emergency ship-to-ship or ship-to-shore communication. Nor was Captain Robert McKay looking for reassurance of any sort when he assumed command of the *Milwaukee.*

~

THE STORM THRASHING Lake Michigan on October 22, 1929, was not your run-of-the-mill November blow. The wind, out of the northeast, ripped the waters at gale force, creating mountainous waves threatening any vessel daring to sail. The shorelines, especially in Wisconsin on the western end of the lake, took a battering that menaced even boats anchored in ports. WORST STORM OF FALL RAGING OVER THE LAKES declared one Wisconsin headline.

But it was more than the worst of the season. Testifying in statements offered over the next couple of months, sailors would proclaim the storm one of the most lethal they had ever confronted. *Milwaukee* purser A. R. Sudon, in the last words he would ever write, declared, "Seas are tremendous. Things look bad."

This was after the *Milwaukee* had already crossed the lake once that day.

When Captain McKay took the eighty-five-mile trip from Michigan to Milwaukee, he knew what his fully loaded boat would be encountering. The storm was already in full force. It would be fair to say, as some did later, that he arrogantly risked life and vessel just to deliver his cargo on time. He knew his vessel and he knew his crew. The *Milwaukee,* despite having an engine with less horsepower than other car carriers, was one of the most powerful boats on the lake, and over the course of his lengthy career McKay had faced dozens of nasty storms. This would be another one.

Except it wasn't. The *Milwaukee* was tormented to her limits, pitching and rolling on the violent seas, the crew members concerned about her cargo. If any of the railroad cars had broken loose, the damage might have been fatal. True to his nickname, McKay took on the heavy weather and emerged victorious. The *Milwaukee* arrived at her destination, where other vessels were already moored, and the crew felt fortunate to be alive. McKay ordered the unloading and reloading of the boat, but few, if any, of his crew believed the boat would make a return trip.

McKay had other ideas. His boat, although taking a beating, had successfully confronted the storm, and he figured to repeat the performance. Car ferries, after all, were known for their strength and endurance. Twenty-seven railroad cars (laden with cheese, lumber, Nash automobiles, and bathtubs and toilet tanks, all valued at $163,500) were carefully loaded onto the *Milwaukee*'s dock tracks, secured in place by a sturdy, complex system that included clamps, holds, and jacks. The cars were coupled the way they would be on railroad tracks on land.

Was the *Milwaukee* overloaded for the sailing conditions? Probably. Under good conditions, the carrier had less freeboard than traditional freighters, especially in the stern, which sat lower in the water, leaving little room between the waterline and the top of the sea gate. In heavy waves, freeboard would be reduced even further. The cars were fully loaded, some with very heavy cargo. In the storm, it would be a major task to keep water from boarding the vessel.

The *Milwaukee* normally sailed with a crew of about fifty-seven men, but on this day, when other captains wisely chose to keep their boats off the lake, five slated crewmembers failed to answer McKay's call to board for departure. Three of the men, reasoning that there was no way they would be sailing under the storm conditions, went downtown to the movies and missed the call. Another flat-out refused to sail. The fifth, according to reports, did not report because he had a bad feeling about the trip.

The *Milwaukee* departed at about 2:30 in the afternoon. A tremendous struggle ensued. Observers onshore said that the boat was fighting a losing battle from the beginning, that her pitching and rolling in the teeth of the nor'easter was unbelievable. These thoughts were echoed by those in the *U.S. Lightship 95*, a lightship anchored three miles from shore. The last to see the *Milwaukee* above water, they said the big carrier was fighting madly to stay afloat. Then she disappeared from their sight.

Something catastrophic occurred a short time later. A large amount of water boarded the vessel and flooded the crew quarters. The sea gate failed, possibly from damage inflicted by railroad cars that had broken loose in the stormy seas. Captain McKay ordered the *Milwaukee* to turn around and head back to shore. Fearing the worst, he ordered his crew to man the lifeboats, with only minimal results.

~

GRAND TRUNK, the U.S. Coast Guard, and other officials wasted no time in acting when the *Milwaukee* failed to arrive in Grand Haven. Another Grand Trunk ferry, the *Grand Rapids*, had left Milwaukee four hours after the *Milwaukee*; when she arrived in Grand Haven, badly battered and late, her captain reported he had seen no sign of the *Milwaukee* during his trip. The storm had devastated land and vessels on both sides of the lake.

Boathouses and small craft were destroyed. Shoreline was wiped out. Bigger boats that had left their docks before storm warnings were posted faced long odds on making it safely home. The *Robert Hobson*, a huge 586-foot steamer built only three years earlier, saw twenty-five thousand of her rivets shear off as the boat, after leaving Indiana Harbor, attempted to sail through the storm; she barely limped into the south Chicago harbor. The *Delos W. Cooke*, a package freighter out of Chicago, warred with the storm for twenty-seven hours before surrendering and returning home. "We just took a long, terrible ride for nothing," her captain, Willis A. Bruso, groused.

There was little doubt that the storm had done in the *Milwaukee*. But what exactly had happened, and where it might have taken place, were questions that no one could answer. It was reasonable to assume that the boat had not sailed far from Milwaukee, but no one knew what to expect. Lake Michigan was humongous, and the storm was strong enough to blow a vessel anywhere. Telegrams flew back and forth between Wisconsin and Michigan. Boats and planes searched nonstop, beginning near the Wisconsin coastline. Everyone hoped to sight lifeboats or survivors. Nothing immediately turned up—no wreckage, no sign of survivors.

By the morning of October 24, the gale had mostly blown itself out and it was easier to survey the lake. The second mate of the *Colonel*, a freighter sailing in southeast Wisconsin, spotted what he determined to be wreckage in the water near Racine. He alerted his commander, Ray Hayward, who ordered the wheelsman in the pilothouse to turn hard starboard. Those on the *Colonel* saw furniture, soggy mattresses, and white-painted wood on the lake surface. Hayward had his crew recover as much as possible and informed Grand Trunk that he might have recovered a portion of the *Milwaukee*'s deckhouse. Planes were dispatched to the scene, but nothing more turned up in the search. The wreckage was clearly from a victim of the storm, but there was no definitive proof that it was from the *Milwaukee*.

Such proof arrived elsewhere. A few miles south, near Kenosha, those on board the steamer *Steel Chemist* recovered two bodies, both wearing life jackets stenciled with SS MILWAUKEE. Other vessels found additional bodies, and it was evident that the *Milwaukee* had sunk. One of those recovered was wearing a watch that had stopped at 9:45.

Confusion over the boat's demise and final resting place increased the following day, when lifeboats were found on the other side of the lake. A Coast Guard team from St. Joseph, Michigan, located a lifeboat, partially submerged in wreckage, containing four sailors. All had perished from exposure. Three were wearing *Milwaukee* life preservers; the fourth, the vessel's wheelsman, was sheltered beneath a tarpaulin.

Two other lifeboats were recovered: one near Holland, Michigan, on the same day as the one with the sailors, and another on October 27, washed ashore with wreckage near South Haven, Michigan. Neither had been used; both were in good condition and fully stocked with untouched supplies. The only conclusion to make was that the boats had been knocked free during the sinking and had ridden the waves to their final resting places.

Some questions were answered in the South Haven wreckage. The *Milwaukee*'s message can bobbed in the water, and when it was recovered, the Coast Guard found a message, written on *Milwaukee* stationery, inside:

> S. S. Milwaukee, October 22, '29. The ship is making water fast. We have turned around and headed for Milwaukee. Pumps are working but sea gate is bent in and can't keep the water out. Flicker is flooded. Seas are tremendous. Things look bad. Crew roll is about the same as on the last payday.
>
> /s/ A. R. Sadan, Perser

Another note, signed "McKay," was also discovered near Michigan. The three-sentence note was nowhere near as specific as Sadon's message:

> This is the worst storm I have ever seen. Can't stay up much longer. Hole in the side of the boat.

The note was difficult to authenticate. McKay's widow examined it, and although she hesitated to call it a fake, she couldn't verify the penmanship as being her husband's. McKay's body had been pulled from the lake, so the note, if a hoax, as many believed, was cruel.

McKay was not around to defend either his reputation or his decision to take his vessel out in the stormy weather. Both received severe criticism in the aftermath of the sinking, and the newspapers, always rushing to judgment after a sinking, were harsh in their assessments. "Was it necessary for the 27-year-old car ferry *Milwaukee* to leave port Tuesday, the day of the severe storm, one of the worst in the history of lake navigation?" asked the *Appleton Post Crescent* in an October 28 editorial titled "Great Lakes Tragedy." The answer to the question was obvious and the writer pushed for accountability. *Somebody* had made the decision to defy Nature. The editorial continued:

> Every fall vessel owners and railroads operating car ferries gamble with death in heavy storms of that period. Every year there are losses of ships and men. Something should be done to curb navigation at this season in the interest of greater safety. The loss of the *Milwaukee* calls for federal investigation, action against those to blame, and the adoption of preventive measures that will be more effective than those now in force.

An investigation conducted by the Steamboat Inspection Service took a measured, methodical approach. There would be no immediate finger-pointing or conjecture, just conclusions based on as many facts as could be gathered.

The first determination involved the *Milwaukee*'s seaworthiness. The vessel had been formally inspected twice in 1929, on June 8 and August 14. Her hull, boilers, and engine had been gone over thoroughly. Minor repairs had been required, but the only major concern centered on a life raft, which was ordered to be replaced. The deck, said Captain Henry Erichson, an inspector stationed in Milwaukee, was in fine overall working order.

So what had sunk the *Milwaukee,* in his opinion?

"I believe her sea gate was carried away," Erichson answered.

The sea gate received a lot of attention during the hearings, mainly because this was the most logical place for water to board the boat. It wasn't the only place, though. The *Pere Marquette 18,* the

Milwaukee's sister ship, had not been equipped with a sea gate, and the car ferry filled with water boarding the boat elsewhere. But this was a place to start. Lee Schuttler, second mate on the *Madison,* testified that he had served on the *Milwaukee* for two years, and during that time he had observed water passing over the sea gate on numerous occasions. The difference between the sea gate of a fully loaded *Madison* and the water was roughly ten feet. The *Milwaukee,* he said, was four feet lower.

Freeboard during the stormy fall shipping season was often a topic of discussion, whether it involved car ferries or freighters. There had been debate over a proposed lowering of weight limits during the late fall season, but nothing had been agreed on when the *Milwaukee* left on her final trip.

There was even a short-lived discussion about Captain McKay's drinking habits. Had he been drinking before the boat departed, and had this affected his judgment? The thought was easily dismissed.

The investigating body would draw only one conclusion: there was insufficient evidence to point to anything specific in the vessel's sinking. It should not have been on the lake during the storm.

〰

THE LOCATION OF THE WRECK remained unknown for decades. Like the *Pere Marquette 18,* it maintained ghost ship status. The *Marquette,* however, was different, as she had been seen before her sinking, and there was at least some idea about where her wreckage was located. The *Milwaukee* might have been anywhere, especially since victims and wreckage had been found on both sides of the lake. She sank long before the invention of sonar scanner devices, so she could have been around any point between Milwaukee and Grand Haven.

In April 1972, the search ended. Fishermen had complained about snagging their net on something, and two amateur sleuths, Kent Bellrichard and John Steele, decided to dive the area and see what they could find. They were about three miles off Whitefish Bay, a northern suburb of Milwaukee.

The *Milwaukee*'s recovered anchor

The *Milwaukee* wreckage

The upright wreckage lay in 122 feet of water. Her wheelhouse was seventy-five feet away. A thorough examination indicated that it was a car ferry, although there was no surefire way of knowing how she had sunk. At least one railroad car had slammed a hole in the side of the boat, but that might have happened during the actual sinking. Oddly enough, the hold chains that secured the cars in rough weather were unused, hanging on their storage racks. Some of the cars were missing, and one was pushed sideways on the deck.

There were no further answers. The *Milwaukee* wreckage, although found, taunted shipwreck historians with more questions.

Wisconsin (1929)

THE WATERS OF LAKE MICHIGAN were just settling down, and the search for the *Milwaukee* shipwreck and her victims was ongoing, when another storm front visited the lake. This one wasn't as violent as the previous one, which had hit with gale-force winds that chewed up property from Milwaukee to Chicago, ruining an ongoing two-year project near Chicago's Lincoln Park, but it did bring down another boat one week to the day after the loss of the *Milwaukee*. This one, the *Wisconsin*, went down near the Wisconsin–Illinois border. Like the *Milwaukee*, she filled with water so quickly as to render her helpless against waves she had no business battling. And like the *Milwaukee*, the *Wisconsin* fought her final fight in darkness, which amplified the chaos of her hopeless struggle. The two vessels even shared a common history with one man: Captain Robert McKay, who died on the *Milwaukee*, had served on the *Wisconsin* in his younger days, when she had been called the *Naomi*.

The *Wisconsin* could claim a colorful history, with many names and owners. Built in 1881 at the Detroit Drydock Company for the Detroit, Grand Haven & Milwaukee Company by Frank E. Kirby, the

Wisconsin and her sister ship, the *Michigan,* were designed as year-round freighters. Kirby spared no expense in making his vessels sturdy and workmanlike, able to crush ice while sailing the route between Grand Haven and Milwaukee. The boat cost a whopping $160,000, but it was well worth the price. What began as hull 49 grew to 204 feet in length, with a beam of 35 feet.

Christened the *Wisconsin,* she sailed for the Goodrich Transit Company until the company lost one of its biggest contracts and sold both her and the *Michigan* to the Detroit, Grand Haven & Milwaukee Railroad, which continued the route. She served admirably in all kinds of weather, taking on winter as designed, until the winter of 1885, a particularly brutal season in which an ice field finished off the *Michigan* and extensively damaged the *Wisconsin.* The vessel was then purchased by the E. G. Crosby firm, which renamed her the *Naomi,* added Muskegon to her schedule, and carried on, without notable event, until May 27, 1907.

The *Wisconsin*

On that spring night, in the middle of the lake with shore in the far distance, the *Naomi* caught fire and looked to be burning to the waterline. Thick black smoke filled the ship and pandemonium might have been the given had it not been for the second mate, John McKay, who, in a loud, clear, and authoritative voice, directed passengers to lifeboats and oversaw their filling and lowering. Three rescue boats, the *Kansas, Saxana,* and *Kerr,* arrived at the site and set about saving all but four of those aboard the stricken vessel. The *Kerr* eased its bow near the *Naomi*'s stern and rescued those passengers unable to get on lifeboats.

The *Naomi* was all but destroyed, but this wasn't the end of the boat. The vessel, sans superstructure, was towed to shore and extensively rebuilt, at a reported cost of $200,000, as a luxurious passenger liner. Renamed the *E. G. Crosby,* she worked as a passenger vessel until the outbreak of World War I, when the government picked her up to act as a hospital boat under yet another name, the *General Robert M. O'Reilly.*

The dizzying array of owners and names ended where it all began. Her original owner, the Goodrich Transportation Company, repurchased her, gave her back her original *Wisconsin* name, and reassigned her to a western Lake Michigan route running between Chicago and Milwaukee. She transported passengers and cargo, and this was her mission in October 1929 when she embarked on her final journey.

~

BY ALL APPEARANCES, Captain Dougal Morrison did not think twice about sailing out onto stormy Lake Michigan on the night of October 28, 1929, despite the death and destruction of the previous week's storm. The *Wisconsin* might have been forty-eight years old, but she was in good shape. Her most recent inspection had been September 17, and she had passed without a hitch.

Loading the cargo could be a time-consuming process, and on this trip, which hinted at rough seas, it was especially urgent that everything belowdecks be as secure as possible. Aside from a small group of passengers, the *Wisconsin* would be carrying an assortment of automobiles, crated machine parts, and iron castings. The crates and barrels were held in place by blocks secured in the hull. The automobiles, brand spanking new, required special attention to be held rigid in what would be a tempestuous lake. The Chicago–Milwaukee trip was going to be rough sailing.

Captain Morrison noticed this as soon as his vessel cleared the breakwater. Most mariners had chosen to stay in on this night, waiting for better weather, the most recent storm still fresh in their minds, but those who did go out would remember pitching and rolling in twenty-foot waves, trying to avoid the troughs between waves, struggling against seas that were boarding at will. It was no different for Captain Morrison. All he had was the knowledge that the trip would be relatively brief.

The *Wisconsin* wrestled with the waves, losing time while it worked northward. The boat was a cacophony of sounds as cargo shifted, occasionally breaking loose and spilling about, the boat's lights flashing into the distance, displaying the horrors that lay ahead. In the engine room, anything not nailed down went flying. Chief engineer Julius Buschmann seemed to be everywhere at once, supervising the crew keeping the boilers fed; making sure that the screw's RPMs were such that the stern, if it lifted from the water, didn't spin out of control and slam back down in the lake, creating a shudder that would rock the boat; and, somehow, in the midst of all the frenetic activity, maintaining the morale of his crew.

Then an almost panic-stricken fireman approached him and shouted that he needed to take a look: a lot of water had entered the boat.

Buschmann and the fireman rushed to the fire hood, where water was rising by the minute. The engineer immediately ordered all pumps to be engaged. Captain Morrison was alerted, and after assessing the situation in the fire hood and being unable to locate the source of the leak, he searched elsewhere on the vessel. The *Wisconsin* was in serious trouble, and if water continued to board the boat it was unlikely that she would make it safely to Milwaukee. If water extinguished the boilers, the engine would shut down and the boat would be without power on a merciless sea.

By that time Morrison was undoubtedly questioning why he had ventured into the storm. Much of his cargo, he noted with alarm, had broken free and was moving unabated with every roll of the vessel. The containers of iron castings had lurched free of their blockings and slammed into the walls of the hold with tremendous force. It would not be long before they pushed out hull plates and added to the leaking. The boat had already developed a list, and this was only the beginning.

The *Wisconsin*'s pumps were unable to keep up with the water filling the boat. Fearing the worst, Morrison went down to the wireless room and instructed Kenneth Carlson to wire for help. The first message was transmitted at about 2:30 a.m.:

> We are four miles off Kenosha. In immediate danger. Please stay with us. May need your help.

The messages became more urgent. The *Illinois,* another Goodrich Transportation vessel originally slated to sail from Milwaukee to Chicago, and therefore in a position to assist the *Wisconsin,* had decided to stay in. The boat's captain said he would head to the *Wisconsin*'s reported position, but it would take time—perhaps too much time—to get there. He wired the Racine and Kenosha Coast Guard stations and explained the *Wisconsin*'s status. The two stations said they would send help.

The transmissions between the *Wisconsin* and the radio marine station grew more frantic. The *Wisconsin* was listing so heavily to her port side that it looked like she might roll over. Morrison did a commendable job keeping his vessel afloat, going so far as to drop anchor to stabilize her as much as possible, but this action doomed the boat. Unable to sail, she was being picked apart by the heavy seas. It was hopeless.

Morrison decided to abandon ship. The Coast Guard had arrived and was prepared to rescue a number of sailors and passengers, but their boats were small and couldn't handle everyone. Larger boats heading to the site were farther away—far enough that the *Wisconsin* was likely to sink before they arrived.

At 4:32 in the morning, three hours after the first transmitted message, the *Wisconsin* sent a notice that painted a grim picture:

> Abandoning ship. Leaving boat now. Can't stay longer. Thanks. Won't forget you.

And, two minutes later:

> Not enough boats for us all.

Men dropped into the icy water; those in lifeboats were picked up by the Coast Guard. Captain Morrison, true to the old maritime tradition, remained on board, fully prepared to go down with his ship.

Nearby, on the Wisconsin shore, a small group prepared to spring into action. People had gathered to watch the ship, in death throes on her side, and among the observers were a lighthouse keeper, Charles E. Young, and his son, Clinton "Tuffy" Young. The *Wisconsin* was only about a mile out, in plain view. They alerted Cliff and Chuck Chambers, owners of the *Chambers Brothers,* a two-year-old 52-foot fishing tug, who had seen the wreck and an overloaded Coast Guard boat bringing in survivors. With several of their friends, they were preparing to brave the waves to offer assistance.

The closer they came to the wreck, the more hellish the scene became. Flotsam was everywhere. Men bobbed in the water, fighting off hypothermia and trying to stay afloat. Some were crying, others praying. Everyone was shouting for help. There were still men aboard the sinking vessel.

Carol Chambers Jornt, niece of the Chambers brothers, recalled the rescue years later. Tuffy Young was somehow able to climb aboard the *Wisconsin* and help men trapped on the boat. "Tuffy got out onto the wreck three times and helped them," she remembered. "If you can imagine anybody with the nerve to do this . . ."

"We spotted twelve men on an eight-man life raft, some so cold they couldn't move," Young said. "I remember that we kept going after the live ones. Finally, we couldn't see any more, so we pulled out of the wreckage and hurried back to the harbor."

Divers salvage the *Wisconsin*'s cargo shortly after the sinking.

The rescue procedures would be debated long after that late October night, when hearings were conducted by the *Wisconsin*'s owners, the Steamboat Inspection Service, and the Kenosha County coroner's office. An official count of those on board was not available, but it is known that fifty passengers and crew members survived, and sixteen, including Dougal Morrison, perished. Morrison was a compelling study of the rescue operation. He had not gone down with the ship; he was pulled from the water alive, only to perish a short time later. Could he—and others—have been saved if the focus had been more on those in the water than on those in lifeboats, who were dry, relatively warm, and in the position to make it to shore without assistance? And who are we to judge the actions of those who risked their lives to help total strangers?

If nothing else, these and other questions illustrate the mass confusion present at any shipwreck. Decisions must be reached instantly, and lives depend on those decisions being the correct ones. Captain Morrison was not alive to explain or defend his decisions, so any conclusions at the hearing were based on conjecture. History would judge and remember Morrison and others based on this conjecture.

Henry Cort
(1934)

THE WIND SHRIEKED and huge waves greeted the *Henry Cort* as she gingerly approached the Muskegon, Michigan, breakwater for a contested turn toward safety on November 28, 1934. The 320-foot steamer, already cursed by three previous sinkings, was taking on one of its most prodigious challenges.

Her day, begun at 9:00 a.m. in Holland, Michigan, not far from Muskegon, had been a trial by anyone's standards. She departed that morning, destined for Chicago and a load of pigiron, and ran headlong into a fierce gale. Most captains prefer to position a vessel's bow into a storm, but the *Henry Cort* was a whaleback, a boat built low to the water and notoriously difficult to handle when taking stormy seas on the bow. The boat's captain, Charles E. Cox, fought the storm admirably until his sailing decisions were taken out of his hands. The storm, tossing the *Henry Cort* about, turned the boat 180 degrees and send her back toward Michigan. Cox decided he might do best to seek harbor in Michigan and restart his journey when weather conditions improved.

That decision, too, was not his to make. Returning to Holland was out of the question, with the heavy seas now behind them. The *Henry*

Cort would have been taking on waves almost directly broadside, and negotiating the breakwater into the harbor was almost suicidal. The other option—heading to Muskegon—wasn't much better. The waves would be more astern, but navigating the breakwater there was going to be touch and go.

Nature had her own designs. Intense currents near the breakwater conspired with the heavy seas there, and the *Henry Cort* was tossed dangerously about. Cox made a last-ditch attempt to stabilize his boat by lowering an anchor, but it didn't work. The *Henry Cort* was lifted out of the water and slammed into the north breakwater. Hull plates ruptured. The bottom of the boat tore when it met the relatively shallow lake bottom. Crewmen were knocked off their feet. The electricity went out, casting the boat into darkness just as water began entering the vessel.

It was a precarious position, to say the least. The *Henry Cort* had come to rest on the breakwater at a thirty-degree angle, her bow jutting heavenward and away from any measure of safety. The shore was three thousand feet away, unreachable in the seas crashing over the breakwater. The men on board the stricken vessel were trapped. And the enormous waves were having their way with the boat.

~

THIS WAS AN INGLORIOUS CONCLUSION for a vessel that had been on the lakes for forty-two years, serving several owners and hauling all kinds of goods, including iron ore and coal. She didn't even start out with her well-known name. When launched in 1892, she was known as the *Pillsbury.*

The whaleback style, also known by the less savory name "pig boat," was the design of Captain Alexander McDougall, a Scottish sailor who moved to the United States, working in Buffalo before eventually settling in Duluth, Minnesota. He sailed on the Great Lakes long enough to know the strengths and weaknesses of an ever-evolving cargo vessel, and he used this knowledge to invent a new, radically different type of freighter—one capable of safely carrying

The *Henry Cort*

more cargo, in all kinds of weather, than the more traditional steamers of the day.

Long, with a cigar-shaped hull and deck housing mounted on turrets, a whaleback would sit low in the water when empty, and lower yet when fully loaded. The flat-bottomed boat had rounded sides that allowed water to wash off it much easier than the freighters that plied their trade on the lakes. The port and starboard sides of the hull joined in the front of the vessel in what looked like a pig's snout, earning the boat its derisive nickname. The more acceptable "whaleback" came from the boat's appearance when sailing: from a distance, it looked like a whale near the water's surface.

The whaleback also boasted one feature that was growing in acceptance in Great Lakes shipping: she was constructed of steel, rather than wood. This gave her the added strength to take on more cargo, a noteworthy characteristic at a time when the Minnesota iron ranges were working overtime to keep up with the increasing demand for steel.

McDougall found the shipping industry averse to his new design. In fact, the whaleback was greeted with scorn by a business reluctant to accept structural change. Undeterred, McDougall used his own money to construct the first whaleback, simply known as the *101*, an unpowered vessel that was towed, laden with cargo, as a barge behind a larger freighter. She was launched on June 23, 1888, and her first journey from Duluth to Cleveland, carrying a load of iron ore, was deemed so successful that it caught the attention of John D. Rockefeller, who agreed to back McDougall. Rockefeller was building an empire in the steel industry, and the whaleback served as a good connection between Minnesota and the big steel-producing companies.

The *Pillsbury,* as the *Henry Cort* was originally called, was not at first intended for hauling ore. Built for the Minneapolis, St. Paul & Buffalo Company and launched on June 25, 1892, she acted as a package freighter, often carrying boxcars of goods to their destinations. Her unique design also assigned her an additional vital function: icebreaker. The Great Lakes' tendency to partially (or, in some years, totally) freeze over made icebreaking vessels an essential part of shipping. Some freighters would have their forepeaks packed and use their size and power to plough through ice and carve out paths for steamers continuing to operate after the usual season. The whaleback was a natural at this. Her structure and weight allowed her to mount and crush the ice in her way.

The *Pillsbury* sailed for almost exactly four years before she was sold to the Bessemer Steamship Company in 1896. Once again, John D. Rockefeller became involved with a whaleback when, in 1901, the Pittsburgh Steamship Company, a division of Rockefeller's U.S. Steel, made her a part of a growing fleet of ore boats. From that point on, the *Henry Cort*'s main cargo was iron ore.

The *Henry Cort* was nearly twenty-five years old and had shipped without serious incident her entire life when the 1917 season began. World War I was taking place, and for the *Henry Cort* and other boats the shipping season was extended, because earlier in the year and later, ore boats were busy hauling ore to support the war effort. The *Henry Cort* and her sister ship, the *James B. Neilson,* formerly the *Washburn,* were heavily engaged in icebreaking.

The *Henry Cort* was a sturdy vessel in adverse conditions, as witnessed in 1913, when a freshwater hurricane passed over the lakes and destroyed boats in its path. The *Henry Cort,* laden with iron ore and, later, a heavy coat of ice, was stronger than other vessels, sinking lower in the water but ultimately sailing safely into port. It was not something the captain and crew, in fear for their lives, would forget.

The *Henry Cort* suffered her first major damage four years later. The winter of 1917 was more brutal than customary on the already frigid waters of the Great Lakes. Thick ice formed on the water, narrowing the rivers that joined the lakes, and with the December temperatures well below zero, boats were in jeopardy of being frozen in until the following spring. The *Henry Cort* and *Neilson* were ordered out as icebreakers by the Pittsburgh Steamship Company and headed up to Bar Point near Lake Erie. Steamers were frozen into the already narrow passage, and a small fleet of freighters were headed in that direction. There was little time to spare.

The two whalebacks arrived and went to work. The open path was narrow, but two of the thirteen freighters made their way through. The *Henry Cort,* captained by John Murray, was going through the tedious forward–backward–forward icebreaking procedure when she was hit by a big freighter, the *Midvale.* The damage, in a congested area of the water, was extensive. The *Henry Cort* was badly holed and sinking fast. All aboard were transferred to the safety of a nearby vessel, but the *Henry Cort* sank in thirty-five feet of water.

Bringing up or salvaging the wreck in such weather conditions was impossible. The *Henry Cort* remained where she was, with only a small portion of her superstructure above water, until the following spring—guarded by the *Neilson,* which kept her from being struck again.

~

NOW, FIFTEEN YEARS LATER, the boat was again in dire straits. Hung up on the breakwater, plunged in darkness, water filling the vessel, the *Henry Cort* appeared to be doomed. Chief engineer Augie Britz, understanding what was happening, had the presence of mind to let the steam out of the boilers before abandoning the engine room. If all went well, there would be no explosion when the lake water hit the boilers.

The initial shock of the accident yielded to a kind of acceptance by those on board that the *Henry Cort* was probably safe for the time being. No one had been seriously hurt when the whaleback ran into the breakwater, and after enduring the indignity of the collision, the *Henry Cort,* though in troubling condition, settled in for the night. The crew believed the help they needed wouldn't arrive until after daybreak; the seas were too rough for a rescue mission. With any luck, the front would pass through, the water would calm somewhat, and the vision that came with daylight would be much better for all concerned.

Much of the crew gathered in the galley, a common practice when times were tough. The collision had been difficult for this area of the ship. Plates, dishes, and utensils had gone flying in every direction. The huge stove and its smokestack were dislodged. The sailors groped about in the darkness. Without power, it was cold.

Things started coming together. Crewmen scrounged around and came up with candles. With minimal light, they began the process of

The *Henry Cort,* heavily damaged

moving the stove and the smokestack back where they belonged. The stove, now functional, gave them heat. The men ate leftovers from Thanksgiving. They eventually broke out some cards and figured they would relax until help arrived in the morning. They were dry, reasonably warm, well fed, and out of the storm.

But help was already on the way. Those on the *Henry Cort* just didn't know it.

A lookout at the Muskegon Coast Guard station had seen the accident, and four men at the station reacted immediately, all knowing that the task ahead would be daunting at the least. The storm had brought down a big boat; they would be manning an underpowered 36-foot surfboat and heading into dreadful conditions. The waters were violent, and, to make matters worse, the pitch-black of night made it impossible to judge the height of the waves the surfboat was navigating. Lifted by the waves and dropped into deadly troughs, the boat hung on.

Nature finally won. A huge wave took the boat completely out of the water, high in the air, and dropped her on her side in the lake. The four men fought for their lives, three making it back to the boat. The fourth, a young man named Jack Dipert, was swallowed by the lake and did not resurface.

THE 1917 SINKING set the *Henry Cort* on a strange trajectory, leading the superstitious to believe the whaleback might be cursed. After twenty-five years of dependable, nondescript service, the *Henry Cort* stumbled through a series of different mishaps and alterations.

She was repaired during the 1918 season but not before she mysteriously disappeared from the scene of her sinking. Pittsburgh Steamship dispatched Captains W. W. Smith, the marine superintendent, and F. A. Bailey, wrecking master, to supervise the *Henry Cort*'s salvaging. When they arrived at the site of the sinking, they discovered the boat had disappeared. The weather during the bad winter season and subsequent spring had moved things around, and with a new shipping season about to commence, it was imperative that the *Henry Cort* be found, lest another vessel find and hit her. A thorough sweep of the area ended on April 23, two weeks into the search, when the *Henry Cort* was located more than two miles from where she had sunk, not far from a shipping lane.

But this was only the opening round of getting her back to functional. Salvagers spent months trying to patch and raise her. They had to repair the hole from the accident and pump out the water from inside the craft, and the process was slow and unsteady. While this costly and time-consuming work was taking place, the *Henry Cort* was found officially responsible for the collision—a huge kick in the teeth, since she had been trying to open the waterway for the boat that hit her. Finally, in late September, she was raised and towed, first to Bar Harbor for temporary work and then to Toledo, where she was overhauled by the Toledo Shipbuilding Company.

The rebuilding was substantial. The *Henry Cort*'s deck had collapsed during the salvage efforts, and the Toledo Shipbuilding Company responded by building a new deck, four feet higher than the original, as well as removing the aft turrets and constructing more traditional (and roomier) quarters in the stern section of the boat. The *Henry Cort* was still a whaleback, but she was now an unconventional one. She lost the entire 1918 shipping season but emerged stronger than ever, ready to take on ice and more cargo.

The *Henry Cort* might have had a new look, but she continued to fulfill her role as an icebreaker. She was, however, wearing out her usefulness on the lakes. She could no longer compete with the giant 500-foot freighters constructed for the iron ore business. She had another mishap in the spring of 1927 when she grounded on Colchester Reef in Lake Erie. The Pittsburgh Steamship Company had seen enough. The firm sold the *Henry Cort* to the Lake Ports Shipping and Navigation Company. The Detroit-based firm specialized in hauling package goods, and the *Henry Cort*'s look changed yet again when she was pulled from the reef and taken to River Rouge, Michigan, for repairs. A more standard deckhouse was installed and cranes were added, which might have worked for the owner's purposes but didn't serve the vessel well. The original McDougall design had been compromised to the point that the whaleback's distinguishing features were no longer noteworthy.

~

THE LOSS OF JACK DIPERT personally bothered John A. Busch, the boatswain's mate in charge of the Muskegon Coast Guard's attempted rescue of the *Henry Cort.* He knew William Dipert, Jack's father, who was stationed just a short distance away at the Point Betsie Coast Guard station and was looking for boats that might have needed assistance. At twenty-three, Jack Dipert had been a young man barely removed from the college football that interested him. The 6-foot-2, 200-pound Dipert had played basketball in high school before studying business administration in college. Like his father, life on the lakes was what he wanted. He'd been with the Coast Guard for five months.

Busch had watched him disappear beneath the waves; the Coast Guard crew fought the fierce waves in an effort to find him but finally gave up. They had to make their way to land or risk losing their own lives. All they could do was hope that Jack had survived and somehow found a way to shore.

Soaked and half-frozen, they were greeted by a growing crowd of people assembled on shore. Word of the *Henry Cort* catastrophe had spread quickly, and people had walked or driven to the site, heedless of the weather. Some had come to assist in any way possible, while others just came to gawk. The dark, hulking shape on the breakwater was unreachable, but additional help was on the way. Busch told his superiors of the *Henry Cort*'s plight and of the failed attempt to rescue anyone on board. They contacted Coast Guard teams in White Lake and Grand Haven, who sent surfboats to the site. The men from Grand Haven moved as close as possible to the wrecked boat and spotted light on the *Henry Cort.* They concluded that at least some of the crew were alive. The vessel appeared to be stable, for the time being.

The *Escanaba*, a Coast Guard cutter that would gain fame during her long life on the lakes, also arrived on the scene. After shining her beam on the *Henry Cort*, her crew drew the same conclusion as the Coast Guard from Grand Haven: there were men alive on the battered vessel. They agreed that any rescue attempts would have to wait until morning, when sunrise would improve visibility.

The forces reassembled early the following morning. Lieutenant Ward Bennett, Tenth Coast Guard District commander, had driven in from Grand Haven the previous evening to supervise. He, too, agreed with the assessment that all rescue attempts should be delayed until daybreak, and he used the intervening hours to coordinate a plan. It would not be anything routine. The waves washing over the stone breakwater had left everything slick with ice and water. The seas, though not as violent as before, still ran rough enough to endanger the surfboats sailing them.

According to the plan, the Muskegon crew, tied together for safety, would slowly make their way on the breakwater until they reached the *Henry Cort*. While they were doing this, a surfboat would move alongside the *Henry Cort* and await further instructions—or, if anyone from the wrecked vessel was washed overboard, rescue them. This sounded like a good idea when it was being discussed, but the waves beat on the surfboat until it finally surrendered and returned to shore.

The Muskegon crew crept very slowly but eventually reached the side of the *Henry Cort*, whose crew anxiously awaited them. There was more work ahead. The wreck towered over the rescue team; they devised a way to bring each crewman down, one at a time, on a makeshift breeches buoy. The men inched their way down until, soaked and freezing, they reached the slippery surface of the breakwater. They then tied themselves to the rescuers and the group began what must have seemed like the longest walk of their lives. Cheering onlookers greeted them onshore. They offered the tired, weather-beaten crewmen hot coffee, but some, victims of exposure, were too cold to accept.

Two of the crewmen, First Mate Harvey Matthews and cook Harry Sutton, were rushed to the hospital for treatment. At sixty-eight, Sutton was in bad shape. He had collapsed during the walk to shore and was carried in a blanket the rest of the way. Matthews had been a leader during the breeches buoy segment of the rescue; he suffered from extreme exposure and likewise required immediate treatment. The remainder of the crew, including Captain Cox, were transported to the nearby Civilian Conservation Corps center, where they received dry clothing and a hot meal.

The *Henry Cort,* battered and broken, remained on the breakwater, a sight for visitors to behold. Cars lined up for miles in the days following the wreck.

To some, the sinking boiled down to a case of good riddance. Chief engineer August Britz, who had been in the engine room during the boat's three final sinkings, had seen enough. "Three times and out," he told the *Muskegon Chronicle* when giving the details of the sinking. "I'm through as a sailor, at least as engineer of the *Henry Cort,* even if the craft is salvaged."

There was no chance of that happening. A pair of December storms finished her off, breaking her in two; the bow section sank, while the stern portion rolled on its side and was quickly broken into pieces. Her life on the lakes was over.

The *Henry Cort* grounded

J. Oswald Boyd (1936)

CARGO WAS ALWAYS FRONT AND CENTER in the commercial shipping business, whether it be furs traded out of birchbark canoes, package goods or lumber in three-masted wooden schooners, or iron ore in enormous steel-hulled freighters. It was always about the profits, even when it involved jeopardizing the safety of sailors moving under less than ideal conditions. Wrecks on the bottom of the lakes bear silent witness to this.

Owners of commercial vessels proved their priorities when they did not insure, or underinsured, their transports while making a point of insuring the cargo. This was clear in the case of the *Edmund Fitzgerald,* the most famous shipwreck in Great Lakes maritime history. The 26,000-plus pounds of taconite was insured; the freighter hauling it was not.

What happened when a vessel was wrecked with its cargo? Maritime laws were fairly specific: if an owner abandoned a wreck, it was essentially a free-for-all.

All this was brought into strange focus in the case of the *J. Oswald Boyd,* a 255-foot steamer transporting just under a million gallons of gas on Lake Michigan in 1936. The *J. Oswald Boyd* grounded on a reef not far from Beaver Island, losing some of her cargo but retaining most of it. The insurance interests wanted no part in the expensive, grueling process of extracting the boat from the reef. The gasoline, precious cargo in the Depression era, became public property, and at least six people died trying to free some of that property for themselves.

IN THE YEARS LEADING UP TO 1936, the *J. Oswald Boyd*'s life had been workmanlike and uneventful, which is what ship owners want from their vessels. Built in Scotland and intended to be an ocean vessel, the *J. Oswald Boyd* was sold to American interests, then sold again to the Gotham Marine Corporation in 1929. She entered Great Lakes service at that time. By all indications, she sailed without incident during her time on the lakes.

Her final run, in early November 1936, was not exceptional. She was to drop off 850,000 gallons of gasoline at the Theisen–Clemens

The *J. Oswald Boyd*

Terminal on the St. Joseph River before heading to Detroit to make a final delivery.

The temperamental weather posed a problem when, on November 8, the *J. Oswald Boyd* encountered a blinding snow squall near the top of Lake Michigan. With little vision, those in the wheelhouse could not make out the usual landmarks they used to guide the boat. The steamer ventured on, but with no direction finder for help, she was sailing without any indication of where she was. She grounded about eighteen miles northeast of Beaver Island, in the shallows known as Simmons Reef, tearing a small but not insignificant hole at the bottom of her forward portside hull. Gasoline began to seep out into Lake Michigan.

All attempts to free the boat met with failure. The captain and nineteen crewmen on board decided to abandon ship; even though it was stable, there was too much danger from the explosive cargo. Three Coast Guard units (Beaver Island, Mackinaw City, and Charlevoix) responded. As it turned out, the water was too shallow for the cutters to draw near the grounded vessel, so, on November 9, the men were moved to Mackinaw City.

And what became of the *J. Oswald Boyd*? Her owner abandoned her to the insurance companies handling the cargo, which was valued at more than $150,000, a considerable sum in those days. Those insuring the vessel offered the gasoline to anyone who wished to remove it. At best, the hole in the boat could be patched and she could be put back into service; otherwise, she could be salvaged for scrap.

The first to visit the *J. Oswald Boyd* were small-timers, mainly men from around Beaver Island, who loaded their boats with barrels or anything else capable of holding the liquid gold. They would siphon the gas out of the *J. Oswald Boyd* and into the containers, bring it back to land, and repeat the process.

One of these men, Peter Nielsen, became the first serious victim of the gas free-for-all. On November 16, Nielsen and a friend had taken Nielsen's tugboat to the grounded vessel and were siphoning gas when the tug's stove ignited and set the smaller vessel aflame.

Burned on the face and hands, Nielsen was rescued from his burning vessel by his friend; before any further damage could be done, the tug was pushed away from the *J. Oswald Boyd.* Their vessel sank, but both men survived.

The Home Insurance Company, a New York firm with the legal claim to the gasoline, caught wind of the Nielsen accident and contracted two salvage companies to remove the rest of the cargo before anyone else was hurt. The insurance company would receive a small portion of profits, on a per-gallon basis, from one of the salvagers; the other salvage firm, the Beaver Island Transit Company, offered to pay a flat $3,700 for what it would remove from the vessel.

This worked for a while. Captain Edward J. Leway, a Cheboygan, Michigan, businessman and sailor, extracted more than twenty thousand gallons, which he took back to Cheboygan and sold wholesale. The Beaver Island Transit Company had a bigger, more sophisticated, and more successful procedure: it armed a wooden package freighter, the *Rambler,* with huge barrels. Using a suction pump mounted on the *J. Oswald Boyd,* the *Rambler* workers were able to transfer 216,000 gallons of gas in less than a month's time.

Winter was settling in, the lake was freezing over, and the *Rambler* was taken off the job for safety considerations. She was replaced

People arrive at the *J. Oswald Boyd* to salvage gasoline.

by a much larger, steel-hulled vessel, the *Marold II,* a onetime luxury yacht and a subchaser during World War I, which Beaver Island Transit hoped would make quicker, more productive work of the task. It ended in disaster. The first trip to the *J. Oswald Boyd* was successful, with eighteen thousand gallons removed. The second voyage, on New Year's Day 1937, was another story.

The day began as a warmer, much calmer day than usual. The men aboard the *Marold* included Everett Cole and Raymond Cole, owners of Beaver Island Transit. The idea was to take advantage of the clement weather before the usual northern Lake Michigan climate made the job more difficult, if not impossible.

All appeared to be running smoothly until late in the afternoon, when residents of Beaver Island and the shore communities in the area heard a thunderous explosion and saw smoke billowing from the *J. Oswald Boyd* wreckage area. A fire was raging, most likely on the *J. Oswald Boyd* and possibly on another boat. The local Coast Guard was alerted and, after finding two vessels burning, called the Charlevoix and Mackinaw Island units for assistance. All boats responding held a respectable distance, but there was no sign of life in the immediate vicinity of the burning boats.

Miraculously, the bulk of the gasoline (in the hold of the *J. Oswald Boyd*) was intact, despite the boat's burning well into the next day. The *Marold II* was a total loss, with its pilothouse, smokestack, and top deck blown onto the *J. Oswald Boyd.* None of the *Marold*'s five-man crew survived. Merrill Cole was found pinned beneath the stack, and his brother was identified by his watch. The hull of the *Marold,* still smoldering the following morning, sank into the lake.

How the *Marold* had exploded will never be known for certain, though plenty of theories abounded. In an area where fumes were thick and there was no shortage of flammable material, speculation was easy. The Coast Guard conducted an investigation and concluded that those on the *Marold* had been negligent of taking the proper safety procedures; after reaching this decision, officials prohibited other boats from visiting the *J. Oswald Boyd* wreckage.

The *Marold II* wreckage strewn atop the *J. Oswald Boyd*

But this did not stop ordinary citizens from ignoring the history and trying to remove gas from the wreckage. The harsh winter converted the water into ice, and people drove for miles to see the site, whether for their share of the booty or just to watch. Hundreds came. On February 28, a man in a truck broke through the ice and drowned. A local dairy used its cans for receptacles—until it discovered that the gas rendered the cans unusable.

~

THE END FOR THE *J. OSWALD BOYD* arrived unceremoniously, which was probably apropos. The boat was freed from the reef in May 1937. It hadn't come easily, but the pilfering of her cargo, along with the removal of water that had entered the vessel over the months, aided in lightening the boat and allowed her, after several attempts, to be freed. She was towed to Michigan, and, ever so slowly, her upper works were sold for scrap.

Her story wasn't quite over. World War II was breaking out, and there was great demand for scrap iron. The *J. Oswald Boyd,* sitting idle for years, was a great candidate. In June 1942, she was towed to the Detroit River and sold to the Great Lakes Steel Corporation. Her metal would be used for the construction of armaments.

The raised *J. Oswald Boyd* wreckage to be salvaged for scrap metal in 1942

William B. Davock (1940)

BY NOW, IT SHOULD BE CLEAR that it would be an understatement to say that storms have wreaked havoc on the Great Lakes. Every year, especially in November, they ravage the lakes. In the old days, before precise measurement of air direction and upper air disturbances, radar, reliable communications between vessels and shore, and other technologies taken for granted today, storms claimed more than their fair share of vessels, including large freighters. Strong Canadian fronts with extremely cold temperatures slammed into warmer southern fronts over the still-warm Great Lakes, causing disturbances that created strong winds, inescapable fronts, deadly waves and troughs, and other conditions that made seafaring difficult, if not impossible.

Historians remember storms of great magnitude—the maelstroms that brought down the *Carl D. Bradley* on Lake Michigan, the *Daniel J. Morrell* on Lake Huron, and the *Edmund Fitzgerald* on Lake Superior. The sinking of vessels in the storms of 1905 and 1913 incurred an inconceivable cost of human life on Lakes Huron and Superior, respectively. The 1913 storm is still considered the rarest of the rare: a hurricane in fresh water.

The storm of 1940, commonly referred to as the Armistice Day Storm because it occurred on November 11, was one of the worst to hit the upper Great Lakes and probably the heaviest storm to torment Lake Michigan. It moved through the area very quickly and decisively, and the only help for the big boats usually out during that time were better weather predictions that kept many freighters in.

Still, the storm took its toll—first in Minnesota, where blizzard conditions quickly followed warmer than average temperatures on a mid-fall day, trapping duck hunters on islands, and later on Lake Michigan, where the storm blew in at a speed and ferocity rarely seen on the lake. Officers and sailors knew of the storm, but not of the speed at which it traveled. In the case of the Armistice Day Storm, this "was the most disturbing day in the history of Lake Michigan."

At first, nothing about the storm appeared out of the ordinary. It started on the West Coast in Washington state, and moved in an almost straight path, west to east, along the U.S.–Canada border. It was cold and dropped a lot of snow (enough to take out a bridge in Washington), but people working the boats were familiar with these conditions.

It gained strength as it moved downward into the Dakotas and Minnesota, where the weather had been unusually good. The cold air behind it brought temperatures down, and the states were socked with snow, heavy as blizzards in some parts. Perhaps with the foresight strong enough to see what lay ahead, farmers strung ropes to their barns so they could find and feed the animals; people were stuck in distress, unable to drive or even catch trains stopped in the bedlam.

The people around the Wisconsin side of Lake Michigan knew about the impending storm but were generally unconcerned. Fall storms usually lasted only a day or two, and they took reports of the conditions in Minnesota. Their boats would wait out the storm and sail in a day or two, when weather improved.

Things were different on the Michigan side of the lake. The vessels there were prepared for an average fall day—sailing that was a bit

Cars stuck during the Armistice Day Storm on November 11, 1940

rough, perhaps, but nothing outside the usual. The sailors would take precautions, seeing that the hatches were battened down securely and cargo secured. No use taking any chances. They had no way of knowing that the storm they were about to confront had gained in intensity to such a degree that it would take down two boats with their crews and ground a third, leaving two sailors lost, in the worst weather conditions ever to face Lake Michigan.

THE *WILLIAM B. DAVOCK,* one of the ships that would be lost with all hands, was built in 1907 by Great Lakes Engineers in St. Clair, Michigan. She was of average length (420 feet) and width (52 feet) for a basic freighter constructed at the time and carried a variety of cargo, from iron ore and stone to grain and coal, from 1907 to 1915. Pickands Mather & Company assumed management of the boat in 1915.

She performed her functions well, moving around the lakes in a regular pattern of delivering coal to ports on Lakes Michigan and Superior and returning to Lakes Erie and Michigan with iron ore she gathered in Minnesota. Contemporary photographs depict her as an attractive carrier with sturdy front and back deckhouses—a vessel well maintained and ready for any kind of travel.

The *William B. Davock*

The *William B. Davock* was reaching middle age when she encountered the 1940 storm. The plan had her heading down the eastern side of Lake Michigan with a load of coal for Chicago. Early weather that morning gave the indication of what boats had to watch for. The daybreak was so nice that one of the porters on the *Charles C. West* woke the first assistant engineer with a rapping on his door. "For gosh sake," the porter cried, "get your camera and come out here to see the prettiest sunrise you ever saw in your life."

The *Charles C. West* was in Milwaukee, headed for southern Lake Michigan. The assistant first engineer, John Schmitt, dressed quickly and went outside. The porter hadn't been exaggerating. The sun refracted shades of red, orange, and purple off the lake. Schmitt happily shot the sights. "It was like being caught in the middle of a giant rainbow and the sun seemed to hang on the edge of the water like a painted backdrop," he later recalled.

Across the lake and much farther north, in the Straits of Mackinac, the crew of the *William B. Davock* got no indication of what stood ahead. The *Henry Steinbrenner,* a much larger carrier that would meet its own ill fate in 1953, followed the *William B. Davock* by about ten minutes. The crews aboard both ships, while hearing of storm weather ahead, saw nothing to indicate that it would be anything but a typical fall storm—enough to rock the boat, but nothing to worry about.

These and other boats went about their business without alarm. One steamer, the 380-foot *Anna C. Minch,* built in 1903 by the American Ship Building Company, had a colorful history of mishaps, but like the others, her crew saw no reason to stay at dock. She had a load of hardwood lumber to deliver, and no autumn storm was going to hold her back.

The *Anna C. Minch,* a Canadian vessel, was traveling from Fort William, Ontario, to Chicago. She was heavily stocked with Canadian sailors; six, including Captain Donald Kennedy, made Collingwood, Ontario, their home, and another seven were from nearby cities. These proximities made for pleasant working relationships, especially on the tedious, often boring trips. Under other circumstances, the Fort William–Chicago route would be one of those journeys.

The storm was picking up in intensity when the *Anna C. Minch* headed out in the early afternoon of November 11. It was still fairly normal sailing, but the waves were something to pay attention to. The *William B. Davock,* sailing four hours ahead of the *Anna C. Minch,* was learning this. By that time, the low-pressure system had moved over the lake with accompanying heavy winds and waves. Other boats, when possible, hightailed it to safety, or at least to a position where they could battle the storm head on.

The *Novadoc* was not one of these boats. Built in Great Britain for Peterson Steamships in Canada, the *Novadoc,* at 253 feet, was a compact vessel designed for the Great Lakes narrows. She was built to haul all kinds of cargo. On this journey, she had a load of powdered coke. If not for a powerful engine, she might have been easy pickings for the storm, whose wind had switched direction and was now coming from the southwest. This mattered. The *Novadoc* had been hugging the eastern shore, trying to hold the lee of 75 mile per hour winds, when she suddenly found herself in great peril. The southwester winds pushed her toward shore, and after several attempts to

The *Novadoc*

fight her power, the ship's master, Captain Donald Steip, decided to try to avoid the troughs, which would likely have brought down the boat—even if it meant grounding her in the sand.

Which is exactly what happened. The boat plowed into a reef near Jupiter Shore, two miles from shore. The force of the grounding began to break up the vessel; ten of the nineteen crew members stayed in the captain's cabin, while the others huddled in the galley. It was early in the evening. With no heat, the crew could only hope for a quick rescue.

It wasn't coming.

THERE WAS NO TELLING what happened to the souls aboard the *William B. Davock* and the *Anna C. Minch,* but there was no doubting that the boats had foundered with all hands. Bodies washed ashore almost immediately in the storm. All kinds of wreckage, from lifeboats to pieces of furniture, including the captain's big wicker chair, made their way to shore. But there was no sign of the boats. They had obviously sunk near one another—that was clear from the men and wreckage recovered—but nothing was found, at least right away.

Elsewhere on the lake, the *Novadoc* hung on, clinging to the hope that the nearby Coast Guard would see them and send help. As it turned out, the Coast Guard could see the stranded vessel, stuck in an area too far from shore for rescue, and had made the unilateral decision, which would be controversial in its aftermath, to wait until the following morning for any kind of rescue attempt. It was hard to say whether anyone on board was still alive. Radio communication had been knocked out by the storm, and from their vantage point, the Coast Guard could see no activity aboard the grounded boat. The storm was stilling. Better to wait until the next morning? The seas may have calmed down by then.

It was a questionable choice by a group sworn to give assistance when needed. Captain A. E. Christofferson, in charge of the Ludington station, had already ordered his watchman off the tower in the afternoon, as visibility was almost completely devastated by snowfall and the wind was so heavy that he legitimately feared the tower might collapse. Sending a vessel to check out the grounded boat was probably suicidal. Communicating with anyone was, at best, spotty. He understood his obligations, but he had never seen something like this in all his years of service. He had responsibilities to himself and his men, and he was going to honor them.

Most of this, of course, was unknown to the officers and crew aboard the *Novadoc,* where there was no heat or food, water seeping in, and nowhere to go. Nor were any of the many people who had heard the quickly spreading word and were now starting to gather on the shore aware of Christofferson's dilemmas and decisions. William Krewell, keeper of the Little Sable Point lighthouse, had seen the *Novadoc*'s masthead lights before she grounded, and while he rushed to inform his superiors about what he had seen, Henry Vavrina, an assistant, went to the boat with a flashlight to let those on the stranded boat know that they knew their plight. He received no indication that anyone on the vessel had noticed.

Conditions were frightful everywhere on the lake. No vessel was safe. Some grounded, some sank, and some limped into port. Their situations were unknown to their companies. With each shipwreck came stories of survivors—not only survivors of the wrecks, though there were plenty of those stories to go around, but people who had avoided certain death by not traveling on a boat to which they would otherwise be assigned. One such story came off the *William B. Davock.* Twenty-four-year-old Duralood S. Farr was a crewman on the boat. He had met and fallen in love with Marcella Jones from Ashtabula, Ohio. The two were to be married, and Farr, already tired of his duties on the *William B. Davock,* decided to sacrifice his annual pay bonus by staying with her. They could have used the additional money, but Farr was determined. Word of the *William B. Davock*'s fate was mixed: Farr was overjoyed by his good fortune, but he lost friends on the boat.

THE STORM DID NOT ABATE the next day. The seas were still high, and the wind remained a problem. Though the intermittent snow was not as heavy as on Armistice Day, it blew everywhere, cutting back vision and making rescue attempts a long shot. Once again, the local Coast Guard decided to stay in, leaving those aboard the *Novadoc* cold, hungry, and wondering about their chances of getting off the grounded boat alive.

Their chances were better than they thought. Three fishermen, Captain Clyde Cross and his two-man crew, Gustave Fisher and Joe Fountana, were aware of the *Novadoc*'s plight and took Cross's boat, the *Three Brothers,* through the heavy seas that might have capsized their craft. They found an ice-encrusted boat, battered and broken, with live sailors on board. Two *Novadoc* crewmen, fearing the worst in back of the ship, had died the day before when they attempted to cross the boat and were swept into the lake by giant waves. All the rest were alive. The men on the *Three Brothers* helped them aboard their craft, bow section first, and took them to shore.

The *Novadoc* shipwreck with onlookers from shore

The *Novadoc* wreckage

The voyage wasn't without issue. Somewhere, somehow, the *Three Brothers* had holed out in the storm. The trip to shore, with the fishing boat loaded with rescued seamen, found the hole dangerously close to the waterline, but the small craft made it. Captain Steip offered Cross financial reward for his vessel and the tense rescue, but Cross would not even consider it.

Cross and his crew were not yet removed from the storm's aftermath. Four days after the *Three Brothers* had been hastily repaired, the men were back on the lake, searching for other boats and survivors. They spotted something sticking out of the water, and after close examination they determined it to be the tip of a boat's mast. The Sarnia Steamship representative found it to be part of the *Anna C. Minch,* in shallow water. Further exploration of the wreck confirmed it to be the boat, with 120 feet of the stern section missing.

The immediate conclusion was that the *Anna C. Minch* had been struck, most probably by the *William B. Davock,* and sank in the storm. After all, men from both boats had been washed ashore. Nothing else turned up, but the theory, largely accepted, was not disproven for more than three decades.

IN MAY 1972, shipwreck hunter Kent Bellrichard finally located the hull of the *William B. Davock* in two hundred feet of water, not far from the remains of the *Anna C. Minch.* Although the water was murky and Bellrichard could barely identify the *William B. Davock,* which had come to rest on her back, he was able to make a positive identification, putting an end to one of the biggest shipwreck mysteries in Lake Michigan history.

The *Anna C. Minch*

Locating the wreck did not end the speculation that the *William B. Davock* had collided with the *Anna C. Minch,* severing her into at least two sections and dropping her to the bottom of the lake. Not enough information had been gathered to disprove the theory.

The story reached its conclusion in early 2015, when the Michigan Shipwreck Research Association, a nonprofit research center based in Holland, Michigan, made public the results of a dive made a year earlier when it sponsored further exploration of the *William B. Davock.* Ironically, nature had provided great assistance to the study. Zebra mollusks, primarily dumped in the water by boats emptying their ballast tanks, had eaten away at the clouding, offering divers a better glimpse of the sunken vessel.

They discovered that the *William B. Davock* had probably been overwhelmed by a sea trough when it was unable to steer in the storm. This disability, the divers proved beyond question, had been caused when the boat's rudder had been jammed against the propeller, breaking off a propeller blade and causing a loss of steering and power. The loss of this vessel and that of the *Anna C. Minch* were unrelated. The coincidence was a huge one, but subsequent dives to the *William B. Davock* indicated no evidence that the boat had struck a thing.

The storm had wreaked havoc on both vessels—separately.

"The waves were so strong they must have snapped the rudder's connection, at which point it swung so far over that it struck the propeller," explained Jeff Voss, who shot the video footage of the wreck. "With no power or steering, the *William B. Davock* would have been at the mercy of the storm."

"It would have sent the ship into what we call a shivering mode," added Craig Rich, an MSRA board member. "It wouldn't have been able to steer, and without the propeller working properly, you can't drive it forward. The ship would have gone into the trough to the wave sideways, which would have swamped the ship."

Valerie Olson van Heest, a scholar, diver, shipwreck hunter, and cofounding member of the MSRA who has spoken of the storm and its effects on the boats sailing in it, may have given the most succinct explanation not only of the 1940 storm but of the effects of storms on Great Lakes shipping. "We must always be vigilant and give Mother Nature respect," she stated. "It doesn't matter if it were one hundred years ago or the present day."

The *William B. Davock* wreckage

NOTES

PHOENIX

Bourrie, *Many a Midnight Ship.*

Ratigan, *Great Lakes Shipwrecks and Survivals.*

Shelak, *Shipwrecks of Lake Michigan.*

Thompson, *Graveyard of the Lakes.*

Captain Edward Carus, "Worst Marine Tragedy in History of Manitowoc Took Toll of 200 Lives When *Phoenix* Burned." Article on file with Wisconsin Historical Society, undated clipping.

NIAGARA

Shelak, *Shipwrecks of Lake Michigan.*

Thompson, *Graveyard of the Lakes.*

Page 23: "boiler explosions . . ." Thompson, *Graveyard,* 35.

Page 24: "It seems . . ." Thompson, *Graveyard,* 95.

LADY ELGIN

Bourrie, *Many a Midnight Ship.*

Boyer, *True Tales of the Great Lakes.*

Ratigan, *Great Lakes Shipwrecks and Survivals.*

Shelak, *Shipwrecks of Lake Michigan.*

Thompson, *Graveyard of the Lakes.*

Van Heest, *Lost and Found.*

Van Heest, *Lost on the Lady Elgin.*

"Fearful Disaster on Lake Michigan," *New York Times,* September 10, 1860.

"The *Lady Elgin* Disaster," *New York Times,* September 13, 1860.

Julian Gurda, "Lost on the *Lady Elgin*: A New Account Emerges," *Milwaukee,* October 27, 2010.

Diane Dombrowski, "Great Lakes Shipwreck Killed 300: A Wisconsin Civil War Story," *Sheboygan Press,* February 19, 2019.

Ron Grossman, "Flashback: 'Waiting for the Waves to Give Up Their Dead': *Lady Elgin* Disaster Sent Hundreds to Their Deaths," *Chicago Tribune,* October 2, 2020.

Page 32: "Queen of the Lakes": Thompson, *Graveyard,* 146.
Page 32: "as a gesture . . ." Shelak, *Shipwrecks,* 87.
Page 36: "I was standing . . ." and "[I] did not notice . . .": "The *Lady Elgin* Disaster."
Page 36: "you could drive . . .": Gurda, "Lost on the *Lady Elgin.*"
Page 38: "I passed through . . .": "Fearful Disaster on Lake Michigan."
Page 40: "We could see . . ." Boyer, *True Tales,* 201.
Page 48: "spread for more . . ." and "the most scattered . . .": Van Heest, *Lost and Found,* 169.
Page 48: "In hindsight . . .": Van Heest, *Lost and Found,* 171.

ALPENA

Bowen, *Shipwrecks of the Lakes.*
Shelak, *Shipwrecks of Lake Michigan.*
Thompson, *Graveyard of the Lakes.*
"The Loss of the *Alpena,*" *New York Times,* October 22, 1880.
Jesse Watkins Jr., "The *Alpena* Mystery," *Chicago Tribune,* October 15, 1950.
Kit Lane, "The Wreck of the *Alpena,*" *Commercial Record,* 1975.
"*Alpena* Went Down October 16, 1880," *Holland Evening Sentinel* [Mich.], October 16, 1980.
Renee Thurber-Prink, "Autumn Has Left Legacy of Ships Lost on Lakes," *Milwaukee Sentinel,* November 2, 1985.
Michigan Shipwreck Research Association, http://michiganshipwreck.org.
Maritime History of the Great Lakes, http://maritimehistoryofthegreatlakes.ca/5825/data?n+97.

Page 49: "like a white swan . . .": Kathy Waines, "Lake Michigan Claims the *Alpena.*"
Page 50: "walking beam": Shelak, *Shipwrecks,* 115.
Page 52: "the worst in Lake Michigan . . .": "Michigan's *Titanic*: The *S.S. Alpena,*" *Michigan in Pictures,* April 14, 2012.
Page 54: "I don't know . . .": "Michigan's *Titanic.*"
Page 56: "with inadequate . . .": "Denial of the Grave Charges Made by the Coroner's Jury," *Cleveland Herald,* January 4, 1881.
Page 57: "unsupported . . .": "Denial of the Grave Charges."

APPOMATTOX

Claire Ried, "Seven Wisconsin Shipwrecks You Can Dive in the Great Lakes," *Milwaukee Journal Sentinel,* n.d.

"Eyes in the Deep: Exploring the Shipwreck *Appomattox*," *Distant Mirror,* July 26, 2011.

Bobby Tanzilo, "On Nov. 2, 1905, the Wooden Steamer *Appomattox* Stranded Off Atwater Beach," [Milwaukee] *Shepherd Express,* November 2, 2021.

R. J. HACKETT

Thompson, *Graveyard of the Lakes.*

James Donahue, "Grounding Snarled Whipping," *Times Herald.* February 25, 1880.

Marinette, November 14, 1905; reprinted in Wisconsin Historical Society, "Steamer *R. J. Hackett*," June 1989.

Page 66: "Lake levels . . .": Donahue, "Grounding."

Page 68: "Just as we left . . .": Wisconsin Historical Society, 49.

Page 70: "great historical . . .": Wisconsin Historical Society, 53.

Page 70: "of considerable appeal . . .": Wisconsin Historical Society, 56.

PERE MARQUETTE 18

Bowen, *Shipwrecks of the Lakes.*

Boyer, *Strange Adventures of the Great Lakes.*

Thomas B. Dancy, "Lake Michigan Mystery: The Sinking of *P. M. 18*," *Telescope,* July–August 1969.

Stonehouse, *Steel on the Bottom.*

"Big Boat Sinks off This Port," *Sheboygan Telegram,* September 12, 1910.

"As Yet No Report on Ferry Wreck," *Sheboygan Telegram,* September 13, 1910.

"Could Save All," *Sheboygan Telegram,* September 14, 1910.

Diane Dembrowski, "'Wrapped Up in Mystery,'" *Herald Times Reporter,* November 28, 2020.

Author interviews with Ken Merryman, Jerry Eliason, and Brandon Balloid

Page 71: "obsession": author interview with Jerry Eliason.

Page 76: "There was only . . .": Stonehouse, *Steel,* 6.

Page 76: "Our perception . . .": author interview with Ken Merryman.

Page 78: "world's greatest . . .": Stonehouse, *Steel,* 9.

Page 78: "nothing more . . .": Stonehouse, *Steel,* 12.

Page 79: "in a tight . . .": author interview with Ken Merryman.

Page 84: "Car ferry 18 . . .": "The Loss of the *P. M. 18,* September 9, 1910." All other citations in this passage are from this source.

Page 85: "When I realized . . .": Dancy, "Lake Michigan Mystery," 91.

Page 86: "The boat was . . .": Dancy, "Lake Michigan Mystery."

ROUSE SIMMONS

Bourrie, *Many a Midnight Ship.*

Boyer, *Ghost Ships of the Great Lakes.*

Lardinois, *Shipwrecks of the Great Lakes.*

Rochelle Pennington, *The Historic Christmas Tree Ship* (Pathways Press, 2004).

Ratigan, *Great Lakes Shipwrecks and Survivals.*

Shelak, *Shipwrecks of Lake Michigan.*

Thompson, *Graveyard of the Lakes.*

"Entire Lake Region Is Swept by Storm," *Milwaukee Sentinel,* December 7, 1912.

"Great Lakes Ships Battered by Storm," *Chicago Record-Herald,* December 7, 1912.

Jay Joslyn, "Documentary Will Shed Light on Yule Tree Ship," *Milwaukee Sentinel,* November 22, 1975.

Joslyn, "Libraries to Get 'Tree Ship' Film," *Milwaukee Sentinel,* May 2, 1984.

"Christmas Tree Ship Honored with Album Commemorating Years on the Great Lakes," [Manitowoc/Two Rivers, Wisconsin] *Herald Times Reporter,* November 7, 1987.

Theodore S. Charney, "The *Rouse Simmons* and the Port of Chicago," *Inland Seas,* Winter 1987.

Dennis McCann, "Load Full of Christmas Trees Went Down with the Ship," *Milwaukee Journal Sentinel,* January 11, 1998.

"The Tale of the 'Christmas Tree' Ship," *Wisconsin Historical Society Newsletter,* November/December 2006.

Glenn V. Longacre, "The Christmas Tree Ship," *Prologue,* Winter 2006.

Suzanne Weiss, "Rogers Street's New Artifacts Building Filed with Fascination," *Herald Times Reporter,* August 7, 2012,

Jordan Tilkens, "Centennial Celebration Commemorates Sunken Ship," *Herald Times Reporter,* November 21, 2012.

John Gurda, "'Christmas Tree Ship' Went Down on the Lake 100 Years Ago," *Milwaukee Journal Sentinel,* December 2, 2012.
Meg Jones, "A Sinking Ship Rises Again," *Milwaukee Journal Sentinel,* n.d.

Page 95: "the gale of . . .": Fred Neuschel, *Anchor News.*
Page 97: "I immediately took . . .": Longacre, "The Christmas Tree Ship."
Page 98: MORE WRECKAGE . . . : *Chicago American,* December 12, 1912.
Page 98: CHRISTMAS SHIP SAFE . . . : *Milwaukee Sentinel,* December 12, 1912.
Page 99: "Everybody goodbye . . .": Bourrie, *Midnight Ship,* 205.
Page 100: "It was a tremendous . . .": Bourrie, *Midnight Ship,* 205.

EASTLAND

Jay Bonansinga, *The Sinking of the Eastland: America's Forgotten Tragedy* (Citadel Press, 2004).
Bourrie, *Many a Midnight Ship.*
Boyer, *True Tales of the Great Lakes.*
Charles River Editors, *The S.S. Eastland Disaster* (Chicago: Charles River Editions, 2023).
Hilton, *Eastland.*
Lardinois, *Shipwrecks of the Great Lakes.*
Jack Silverstein, "100 Years Later, the Jury in *Eastland* Disaster," *Chicago Daily Law Bulletin,* June 19, 2015.
Shelak, *Shipwrecks of Lake Michigan.*
Susan Q. Stranahan, "The *Eastland* Disaster Killed More Passengers Than the *Titanic* and the *Lusitania.* Why Has It Been Forgotten?" *Smithsonian,* October 2014.
Sutton, *Capsized!.*
Thompson, *Graveyard of the Lakes.*
"*Eastland* Disaster as Reporter Saw It," *New York Times,* July 25, 1915.
"Stretchers Made an Endless Chain," *New York Times,* July 25, 1915.
"Little Feller Now Has a Name," *Chicago Tribune,* July 30, 1915.

Page 108: "In the *Eastland* . . .": Michael McCarthy, *The SS Eastland,* reprinted in *The Eastland Disaster,* unpaginated.
Page 109: "The boat kept . . .": Charles River Editors, *S.S. Eastland,* n.p.
Page 110: "[He] stood on . . .": Charles River Editors, *S.S. Eastland,* n.p.
Page 113: "People were packed . . .": Boyer, *True Tales,* 43.

Page 114: "919 Bodies . . .": Boyer, *True Tales,* 44.
Page 115: "Who is . . .": Boyer, *True Tales,* 46.
Page 116: "The hearts . . .": Charles River Editors, *S.S. Eastland.*
Page 117: "While still . . .": memorial plaque for the *Eastland,* posted on the Chicago River.

MILWAUKEE

Boyer, *Ghost Ships of the Great Lakes.*
Lardinois, *Shipwrecks of the Great Lakes.*
Ratigan, *Great Lakes Shipwrecks and Survivals.*
Shelak, *Shipwrecks of Lake Michigan.*
Thompson, *Graveyard of the Lakes.*
"Car Ferry *Milwaukee* Sinks with All Crew: Bodies Washed Ashore," *Escanaba* [Mich.] *Daily Press,* October 25, 1929.
Scottie Dayton, "Great Lakes Car Ferry *Milwaukee* (1929)," *Ship in Scale,* July/August 1990.
Jim Hettinger, "'Heavy Weather' McKay's Fatal Error," *Enquirer,* December 26, 2015.
Jarod Riemersma, "CG Search for *SS Milwaukee,* History of the Upper Great Lakes," n.d.
Bobby Tanzilo, "On October 22, 1929, the Grand Trunk Car Ferry *Milwaukee* Sank in a Storm," *On Milwaukee,* October 22, 2020.
Steamboat Inspection Service to Supervising Inspection Steamboats, Detroit, Michigan, October 24, 1929.
Steamboat Inspection Service to Steamboat Inspection, Milwaukee, Wisconsin, October 25, 1929.
Steamboat Inspection Service to Supervising Inspection Steamboats, Detroit, Michigan, October 25, 1929.
Steamboat Inspection Service to Supervising Inspection Steamboats, Detroit, Michigan, October 26, 1929.
Steamboat Inspection Service to Local Inspection Steamboats, Milwaukee, Wisconsin, October 30, 1929.
Local Inspection Service, Milwaukee, Wisconsin, to Steamboat Inspection Service, October 31, 1929.
Handwritten and typed notes regarding the accident from the United States Coast Guard, December 26, 1929.
News, October 22, 1929.

Page 122: "seas are tremendous . . .": Boyer, *Ghost Ships,* 91.
Page 126: "*S. S. Milwaukee* . . .": handwritten notes from the Coast Guard, December 26, 1929.
Page 127: "Was it necessary . . .": "Great Lakes Tragedy," *Appleton Post-Crescent,* October 28, 1929.
Page 127: "I believe . . .": Boyer, *Ghost Ships,* 97.

WISCONSIN

Boyer, *Ghost Ships of the Great Lakes.*
Shelak, *Shipwrecks of Lake Michigan.*
Joseph J. Jacoby, "1929 Sinking of *S.S. Wisconsin* Stirs Many Local Memories," *Kenosha News,* n.d.
Joseph J. Jacoby, "Scuba Divers Search Sunken Ship," *Kenosha News,* November 23, 1962.
Leonard Palmer and Diane Giles, "Black October," *Kenosha News,* October 23, 1984.
Diane Giles, "Shipwrecked," *Kenosha News,* October 30, 2005.

Page 135: "We are four miles . . .": Boyer, *Ghost Ships,* 107.
Page 136: "Abandoning ship . . .": Boyer, *Ghost Ships,* 107.
Page 136: "Not enough boats . . .": Boyer, *Ghost Ships,* 107.
Page 137: "Tuffy got out . . .": Boyer, *Ghost Ships,* 107.
Page 137: "We spotted twelve . . .": Boyer, *Ghost Ships,* 107.

HENRY CORT

Boyer, *Strange Adventures of the Great Lakes.*
Passwater, *Historic Shipwrecks and Rescues on Lake Michigan.*
"*Henry Cort,*" Michigan Shipwreck Research Association, n.d.
Eric Gaertner, "*The Chronicle* Reported *Henry Cort*'s 25-Man Crew Lost," *Muskegon Chronicle,* November 27, 2009.
Eric Gaertner, "75 Years Later, the Sinking of the Steamer *Henry Cort* Remembered," *Muskegon Chronicle,* November 27, 2012.
Eric Gaertner, "West Michigan Underwater Preserve Officially Recognized by State to Promote, Protect Shipwrecks," *Muskegon Chronicle,* November 27, 2012.
Dave LeMieux, "Lookback: *Henry W. Cort* Freighter Involved in Fatal Accidents Later Meets Own Demise," *Muskegon Chronicle,* January 2, 2014.

J. OSWALD BOYD

Ratigan, *Great Lakes Shipwrecks and Survivals.*
Shelak, *Shipwrecks of Lake Michigan.*
Le Roy G. Barnett, "A Fuel Free-for-All," *Michigan History,* March/April 2023.
"The *J. Oswald Boyd,*" December 21, 1943 (court case notes).

WILLIAM B. DAVOCK

Page 162: "For gosh sake . . .": Boyer, *Strange Adventures of the Great Lakes,* 75.
Page 162: "It was like . . .": Boyer, *Strange Adventures of the Great Lakes,* 80.

BIBLIOGRAPHY

Bourrie, Mark. *Many a Midnight Ship.* Ann Arbor: University of Michigan Press, 2005.

Bowen, Dana Thomas. *Shipwrecks of the Lakes.* Cleveland: Freshwater Press, 1952.

Boyer, Dwight. *Ghost Ships of the Great Lakes.* Cleveland: Freshwater Press, 1968.

Boyer, Dwight. *Great Stories of the Great Lakes.* Cleveland: Freshwater Press, 1966.

Boyer, Dwight. *Strange Adventures of the Great Lakes.* Cleveland: Freshwater Press, 1974.

Boyer, Dwight. *True Tales of the Great Lakes.* New York: Dodd, Mead & Company, 1971.

Green, Catherine M., Jefferson J. Gray, and Bobbie Malone. *Great Ships on the Great Lakes. Wisconsin Historical Society Press. 2013.*

Hilton, George W. *Eastland: Legacy of the Titanic.* Stanford: Stanford University Press, 1995.

Hollander, M. Paul. *Lost Lady: The Lady Elgin Tragedy.* Self-published, 2018.

Lardinois, Anna. *Shipwrecks of the Great Lakes: Tragedies and Legacies of the Great Lakes.* Lanham, Md.: Globe Pequot, 2021.

Oleszewski, Wes. *Ghost Ships, Gales and Forgotten Tales.* Marquette, Mich.: Avery Color Studios, 1995.

Oleszewski, Wes. *Mysteries and Histories: Shipwrecks of the Great Lakes.* Gwinn, Mich.: Avery Color Studios, 1997.

Oleszewski, Wes. *Shipwrecks and Rescues: Forgotten Great Lakes History.* Gwinn, Mich.: Avery Color Studios, 2008.

Passwater, Michael. *Historic Shipwrecks and Rescues on Lake Michigan.* Charleston: History Press, 2022.

Ratigan, William. *Great Lakes Shipwrecks and Survivals.* Grand Rapids, Mich.: William Eerdmans Publishing, 1960.

Schumacher, Michael. *Too Much Sea for Their Decks: Shipwrecks of Minnesota's North Shore and Isle Royale.* Minneapolis: University of Minnesota Press, 2023.

Schumacher, Michael. *Wreck of the Carl D.* New York: Bloomsbury, 2008.

Shelak, Benjamin J. *Shipwrecks of Lake Michigan.* Black Earth, Wisc.: Trails Books, 2003.

Stonehouse, Frederick. *Steel on the Bottom.* Gwinn, Mich.: Avery Color Studios, 2006.

Sutton, Patricia. *Capsized! The Forgotten Story of the SS Eastland Disaster.* Chicago: Chicago Review Press, 2018.

Swayze, David D. *Shipwreck!* Boyne City, Mich.: Harbor House, 1992.

Textor, John. *Phoenix: The Fateful Journey.* Sheboygan, Wisc.: Sanderling Press, 2006.

Thompson, Mark L. *Graveyard of the Lakes.* Detroit: Wayne State University Press, 2000.

Van Heest, V. O. *Lost and Found: Legendary Lake Michigan Shipwrecks.* In-Depth Editions, 2012.

Van Heest, V. O. *Lost on the Lady Elgin.* In-Depth Editions, 2010.

ILLUSTRATION CREDITS

The University of Minnesota Press gratefully acknowledges the following institutions and individuals who provided permission to reproduce the illustrations in this book.

Pages 3, 31, 39, 62, 81, 129 (lower image), 137: Wisconsin Historical Society.
Pages 4, 157, 167 (upper image): Alpena County George N. Fletcher Public Library, Alpena, Michigan.
Page 6: Hope College Library, Van Ark Postcard Collection, Myron Van Ark via the Joint Archives of Holland, Archives and Special Collections.
Page 7: DN-0061476, *Chicago Daily News* collection, Chicago History Museum.
Pages 11, 26, 69, 70, 91: C. Patrick Labadie Collection, Thunder Bay National Marine Sanctuary, Alpena, Michigan.
Pages 15, 144, 149, 170: Michigan Shipwreck Research Association.
Page 22: Photograph by Keith L. Courtesy of the Historical Marker Database (hmdb.org).
Pages 37 and 41: *Frank Leslie's Illustrated Newspaper,* September 22, 1860, Brendon Baillod Collection.
Page 40: The Charles Deering McCormick Library of Special Collections and University Archives, Northwestern University Libraries.
Pages 50, 85, 105, 109, 152, 154, 167 (lower image): Kenneth Thro Collection, University of Wisconsin–Superior.
Page 53: Fr. Edward J. Dowling, S. J. Marine Historical Collection, University of Detroit Mercy.
Pages 61, 164, 168: Historical Collections of the Great Lakes, Bowling Green State University.
Pages 65, 90, 162: Lake Superior Maritime Collections, University of Wisconsin–Superior.
Page 75: Detroit Publishing Company [between 1900 and 1910]. LC-D4-34721 [P&P], Library of Congress Prints and Photographs Division, Washington, D.C. (http://hdl.loc.gov/loc.pnp/pp.print).
Pages 93 and 94: Chicago History Museum.
Page 101: Hubert R. Wentorf, 1971. Courtesy of Lester Public Library, Two Rivers, Wisconsin.

Page 104: Detroit Publishing Company [1911?]. LC-D4-22668 [P&P], Library of Congress Prints and Photographs Division, Washington, D.C. (http://hdl.loc.gov/loc.pnp/pp.print).

Pages 111 and 117: The Newberry Digital Collections.

Page 112: DN-0064948, *Chicago Daily News* collection, Chicago History Museum.

Pages 121 and 132: Great Lakes Marine Collection of the Milwaukee Public Library and Wisconsin Marine Historical Society.

Page 129 (upper image): Photograph by Paul F. Courtesy of the Historical Marker Database (hmdb.org).

Page 141: Detroit Publishing Company [photograph collection]. LC-D4-22839 [P&P], Library of Congress Prints and Photographs Division, Washington, D.C. (http://hdl.loc.gov/loc.pnp/pp.print).

Page 154: Beaver Island Historical Society Archives.

Page 161: Minnesota Historical Society.

MICHAEL SCHUMACHER is the author of seven books about Great Lakes shipwrecks, including five published by the University of Minnesota Press. He has written narratives for twenty-five documentaries on Great Lakes shipwrecks and lighthouses. He lives in Wisconsin.